CLIPPING
HORSES & PONIES

CLIPPING
HORSES & PONIES
A Complete Illustrated Manual

SHARON LLOYD

KENILWORTH PRESS

First published in Great Britain 1995 by
The Kenilworth Press Ltd
Addington
Buckingham
MK18 2JR

© Sharon Lloyd 1995

British Library Cataloguing in Publication Data
A CIP record for this book is available from the British Library

ISBN 1-872082-77-7

Cover design by Paul Saunders
Cover photos by John Birt
Typeset in 10/13 Plantin
Design, typesetting and layout by The Kenilworth Press Ltd
Printed and bound in Great Britain by
Butler and Tanner Ltd, Frome

....

*This book is dedicated to all the horses who have taught me so much over the years,
especially my pony Cefnllan Copper.*

And also to William who is still teaching me.

•CONTENTS•

ILLUSTRATION CREDITS

Photographs

All photographs are by *John Birt*, except the following:

Aerborn Equestrian Ltd: 92 (bottom left) – Aerborn Husky New Zealand; 94 – Aerborn waterproof exercise sheet.

Bossy's Bibs: 92 (top right).

Chris Cook, Pleasure Prints: 36.

Cox Surgical Ltd: 106 – Wahl battery trimmer and Sterling 2 Plus rechargeable trimmer.

John Henderson: 34.

Horse & Rider: 110, 111, 112, 113.

Bob Kitchen: 12 (right).

Lister Shearing Equipment Ltd: 103 (left) – Stablemate clipper; 103 (right) – Laser clipper; 107 – Lister blades.

Sharon Lloyd: 11 (bottom), 14, 16, 17, 18, 22, 24, 41, 45, 46, 92 (top left).

Anthony Reynolds: 10 (bottom), 11 (top), 12 (left), 30.

Mike Roberts: 10 (top).

Weatherbeeta Ltd: 89 (top) – Weatherbeeta 300 Superquilt; 89 (bottom) – Gold Line Thermaquilt; 91 (left) – 'Gusset Extra Deep' New Zealand; 91 (right) – Weatherbeeta 700; 93 – Weatherbeeta hood and neck cover; 105 (bottom) – Sunbeam-Oster clippers.

Line illustrations

The line illustrations are by Kenilworth Press, except for those on pages 110 and 111 which are reproduced by kind permission of Lister Shearing Equipment Ltd.

•ACKNOWLEDGEMENTS•

Many thanks to all the following companies for their help, especially Alun Williams of Lister Shearing Equipment Ltd for his invaluable contribution in providing information, photographs and equipment for this book; to Weatherbeeta Ltd and Aerborn Equestrian Ltd for supplying photographs of their superb rugs for the management section; to Bossy's Bibs and Cox Surgical for supplying photos; *Horse & Rider* magazine, for allowing us to use their excellent pictures showing clipper maintenance.

To Helen Owen and her two beautiful dressage horses Henry and René, who were very patient models for the clipping photograph session at Helen's yard; and to John Birt for taking the photographs there.

Thanks to Judy Howard at The Dr Edward Bach Centre and everyone at the Radionic Association especially the School of Radionics for all their help and support during my training.

To Lorna Walters for her valuable suggestions on the text.

Special thanks to Mike Pegington for looking after my horses while I went galavanting around the country putting this book together and away on various courses. Also to Angela West and Breeze, Jeanette, Jennie and Kirsty McCarthy with Saz and Star Wars, Kate Hewitt and Bally for being models for some of the photographs, and everyone at Pentwyn Barn Farm, Llangwm, Gwent.

Thanks also to Clive Morgan, my ever-prompt blacksmith, Nigel Alred of Alred's Agricultural Services, for catering for my horses' nutritional requirements for many years, and all at Boughsprings Farm Shop, Ponthir, Gwent.

Many thanks to Ren Pike for helping with the electrical safety section and for keeping my very ancient set of Lister Showman clippers running.

And last but not least to Gillian McCarthy BSc (Hons), MBIAC, for contributing to the book and without whose constant nagging this book probably would never have been written.

Sharon Lloyd
August 1995

• 1 •

WHY AND WHEN TO CLIP

Winter can be a frustrating time for many horse owners, especially if they dread the thought of having to have their horses or ponies clipped, or of doing the clipping themselves. Clipping unwilling horses can be a nightmare for both horse and human, but it does not always have to be like this. With good preparation, practice and a little 'horse sense', the process of clipping may not be as problematic as first imagined.

SO WHY CLIP?

Equines have evolved over millions of years and have adapted very well to a variety of environments and climates. Even the diverse weather conditions from season to season remind us of how the horse can quite easily make allowances for temperature and atmospheric changes. In Northern Europe, by the end of October most horses and ponies have grown a full winter coat which is thicker and longer than the short sleek summer coat, and this is in readiness to protect themselves against the cold and wet weather that is inevitable. This is nature's way of preventing horses and ponies from losing condition, and keeping them healthy and happy, by attempting to compensate for the bad weather and sparse amount of food of poor nutritional value that is available in their natural environment during the winter months.

However, with the increasing trend of keeping horses stabled, out of bad weather, and providing them with concentrated food and artificial warmth by rugging, this natural protection isn't so necessary. In fact, for horses expected to work (be exercised, hunt or compete) during the winter months, it can be positively harmful, as outlined below.

ADVANTAGES OF CLIPPING

❍ A horse with a thick winter coat will overheat quite quickly with fast work or mild winter weather and sweat excessively to compensate. This is very unhealthy for a variety of reasons. Firstly,

Native horses and ponies have evolved over time to adapt to the British climate. Even the summers can be very cold on exposed moors and mountains so protection against the climate is just as important at this time of year.

it can cause a great deal of distress during work. Secondly, if the horse is worked fairly hard every day, then over a short period of time the horse will lose condition, even if you think he is being adequately fed to compensate for the energy used. Clipping will help a horse stay cooler during work, so preventing excessive sweating, and he can work for longer periods without becoming overheated, stressed and tired.

❍ Because of the excess sweating and harm that can be done by working a horse for any length of time with a full winter coat, it is virtually impossible to get such a horse fit enough for even a novice competition, so clipping is essential if any standard of fitness is required.

❍ A thick winter coat will take hours to dry off properly after exercise and care must be taken not to allow the horse to stand around for hours with a wet coat, without being rubbed down and rugged to keep warm. If this happens, a horse

will quickly become very cold and catch a chill that could eventually lead to pneumonia. A horse that has been clipped will cool down sooner and dry off much quicker after being worked.

❍ It is very difficult to groom a horse thoroughly when the winter coat is fully grown. Dried sweat cannot be removed properly and sores can result from tack or rugs rubbing the horse. Also, waste products produced through sweating, if not removed, can clog the pores and this may result in skin infections. Clipping does make the task much easier but should never be a short cut for thorough grooming.

❍ Clipped horses are easier to keep clean, especially those who like to 'wallow' in mud when turned out.

Clipping plays a part in producing show horses, as seen with this handsome show cob with a hogged mane and trimmed legs.

A pony with this amount of coat will certainly need a clip if he is to work hard.

○ Parasites love the warmth and security that a long coat provides during the winter. If clipped, the horse is less likely to suffer from these. If, however, the horse does pick up parasites, then the appropriate treatment, such as with powders or washes, will be more effective on a shorter coat.

○ Sometimes your vet will advise you to clip the coat, either all over or in a specific area to facilitate treatment of a skin condition. If the horse has to be washed frequently with lotions etc., drying off after treatment is much easier. After an injury, your vet may clip a small area around the wound to help keep it clean and make stitching easier. This is usually done with very fine blades. If pus seeps out of the wound, then the short hair

allows it to be easily washed off, likewise any bedding and dirt that sticks to the wound. Clipping the injury site seems to help cuts and wounds to heal faster and may in some cases prevent scarring because the wound edge is tidier. However, if some scarring is inevitable, the scar usually has a much 'neater' appearance.

WHEN TO CLIP

In Great Britain, the winter coat usually starts to grow around mid to late September. The growth of the coat is influenced by temperature, food and the amount of sunlight available. The first clip is not usually done until the winter coat is well established, but the timing can vary with individual horses depending on breed, work requirements, condition and management.

October is the traditional time for clipping in Britain. This has come about because hunters need to be both fit and clipped in time for the opening meet of

Native-type pony with full thick winter coat.

Clipping is necessary in the winter months so that horses can continue to do fast work and demanding activities such as these without becoming over-heated and distressed because of a thick winter coat.

the hunt, which is usually in late October or early November. These horses are therefore clipped in early October so that they can be worked properly in preparation. Depending on the rate at which the coat grows back, the second clip is done around Christmas. If another clip is required, this is usually done in early February.

Most textbooks say that horses should not be clipped before October because the winter coat is not fully grown, nor after February as it will interfere with the growth of the new summer coat. Sometimes, though, it is necessary to clip outside these recommendations to safeguard the comfort and well-being of individual horses. For example, if the weather in late August and early September is particularly bad, then the winter coat may grow in very quickly, and if the horse continues to do regular

fast work, then each time he does so he will become very sweaty and gradually lose weight. This is at a time when he needs to be in very good condition in preparation for the winter months, so in this case clipping earlier would be essential to prevent the horse losing valuable body weight.

MAKING THE DECISION TO CLIP

Don't ever rush into clipping. Before any definite plans are made, first ask yourself whether clipping is really necessary. Remember that horses are clipped for a specific reason. Clipping should not be done for cosmetic purposes, fashion or the fact that everyone else in your yard or in the locality is doing it. Ask yourself: Is the work going to be often enough or hard enough to warrant the horse being clipped? Clipped horses require just as much, and in some instances, more time, money and care than unclipped horses – so think before you clip!

If, for whatever reason, you feel that clipping is not going to be feasible, advice on how to avoid it altogether is given at the end of Chapter 4.

·2·

DECIDING ON A CLIP

TYPES OF CLIP

There are a variety of clips to choose from but, before any decisions are made, there are many things to take into consideration. It is advisable to think about the management and the work requirements of the horse through the winter. The choice of clip should be geared to the individual needs of the horse. To clip off more coat than required is very bad management and also very uneconomical as the cost of keeping your horse is increased as he will need extra feed, hay, bedding and rugs to make sure he is warm and happy.

The main places that become sweaty and uncomfortable when the horse is doing hard work are: the jowl region, the underside of the neck to between the front legs, the girth area and around the elbows, between the back legs, the flanks and especially the saddle area. If the horse is in very hard work, such as competitions, hunting or racing, then most of the coat will become saturated so a large area of the coat may need removing for the horse to remain comfortable.

NECK AND BELLY CLIP

This clip is ideal for horses and ponies who are living out through the winter and only used for light hacking at the weekends or one or two evenings during the week. The coat is taken off just under the neck and along the underside of the belly. Stabled horses who feel the cold or

An apron clip – for horses and ponies in light work.

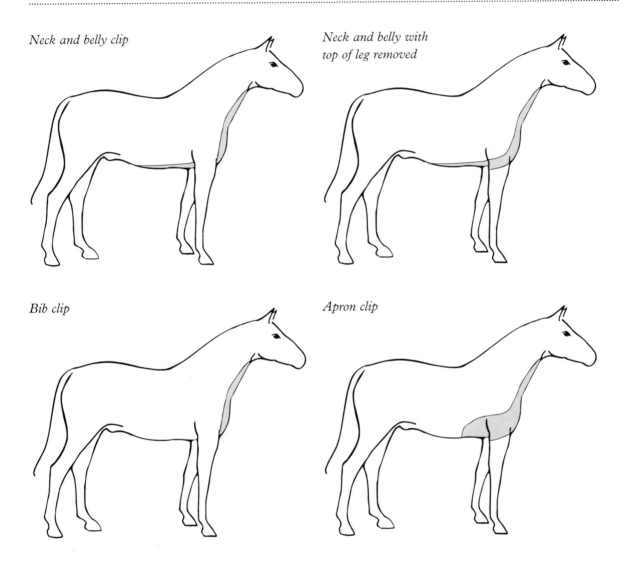

Neck and belly clip

Neck and belly with top of leg removed

Bib clip

Apron clip

are doing light work can benefit greatly from this clip. Field-kept and stabled horses will still need to be rugged. A good quality New Zealand rug should be used for the horse living out and also for the stable-kept horse who is turned out for a few hours during the day.

If you do not like the thought of your horse lying down in wet mud without any protection on the belly then there are variations to this clip. One is the **'bib'** clip, which consists of removing the coat from the underside of the neck down in front of the chest. The other is the **'apron'** clip, which takes off more coat to the girth line between the front legs and the top of the forelegs. Extremely tough native ponies don't usually need a rug with a bib or apron clip, but common sense should prevail. Keep a close eye on the condition of the pony and on what is happening with the weather. Take appropriate action if either deteriorates.

An Irish clip on a pony who is in light work and turned out during the day with a New Zealand.

belly clip but not enough to warrant a more drastic clip. The clip usually finishes at the jowl but sometimes hair from the lower part of the face is removed. Again, it is beneficial for stabled horses who feel the cold, and because the coat is left on the hind legs and thighs, field-kept horses are protected from the cold wind.

When horses are out in the field and the weather is wet and windy they will always turn their quarters into the wind. This way the more vulnerable front end of the horse, including the head and neck, is protected from losing heat. If the coat between the buttocks, around the hind legs and either side of the tail is removed, a great deal of heat will be lost because the skin is much thinner there. This clip therefore offers some protection. It is also advisable not to thin the tail or pull the top of the tail on a clipped horse who will be living out or turned out for long periods in the daytime. A full tail will help to act as a windbreak and heat insulator. Rugs are also needed: a stable rug for the horse living in, and a New Zealand for the horse living out and the

THE IRISH CLIP

The Irish clip is very similar to the neck and belly clip described above, but more coat is taken off the chest, belly and shoulders giving a 'triangle' appearance. This clip is useful for both stabled and field-kept horses who are doing slightly more work that those needing a neck and

Irish clip

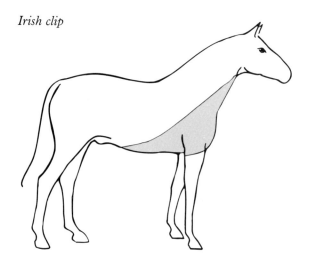

Irish clip with half head removed

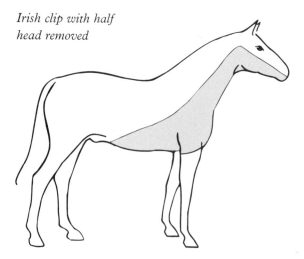

A low trace clip.

stabled horse who is turned out.

An Irish clip is handy for horses who kick, as the rear end of the horse is left untouched (beware of 'cow-kickers' though!). Interestingly this clip also gives the illusion of making the horse appear to support itself and be more 'up in front', which is especially handy if it tends to have most of its weight on the forehand while working. However, this clip should not be used to compensate visually for incorrect riding and schooling!

LOW TRACE CLIP

This clip is also excellent for field-kept and stabled horses and ponies who tend to sweat when exercised but are not going to be doing any hard, fast work or competing. The hair is removed from the underside of the neck and belly, between the forelegs and the upper part of the

hind legs. The lower half of the face can also be removed. As a guide to the amount of hair to be removed from the belly, put the horse's saddle on and measure roughly 7 inches (approx. 18cm) from the bottom of the saddle flap. Again, rugs should be used on horses with this clip to compensate for the coat that has been removed from the neck and belly.

Low trace clip

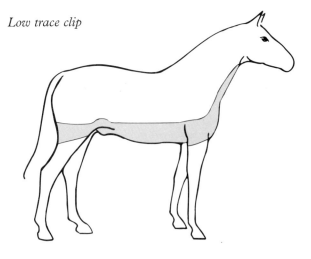

The name of this clip originates from harness horses – the height of the clip was always level with the height of the traces. This style of clip is now widely used on riding horses. The appearance of horses with short backs can be improved with the trace clip, as they will look more in proportion.

MEDIUM TRACE CLIP

This clip is more or less the same as the low trace clip but more of the coat is clipped off. Horses living out could still have this clip but great care must be taken with regard to feeding and warmth. If this much coat is removed and the horse or pony is still living out, then a field shelter is essential. (Personally, I would prefer horses with this clip to be stabled.) Although there is still quite a lot of hair remaining to give

natural protection, in very cold, wet weather it may be necessary to add an under-rug to the New Zealand or stitch another blanket onto the lining. The under-rug could be put on at night and removed during the day, if the days are mild. Stabled horses will also need good quality rugs and a New Zealand if turned out in the day.

Horses who are in regular work and may be required to go on faster rides or do the occasional 'fun ride' would probably require this type of clip. As a good guide to the height of the clip, take off the hair just below the bottom of the saddle flap.

If stabled, then the whole or half of the head can be clipped for a tidier appearance. If half the head is clipped then the line should follow the path of the cheekpieces down the face. Some thought should be given as to whether the head should be clipped, as this part of the

A medium trace clip with half the head removed.

Medium trace clip

High trace clip with half a head

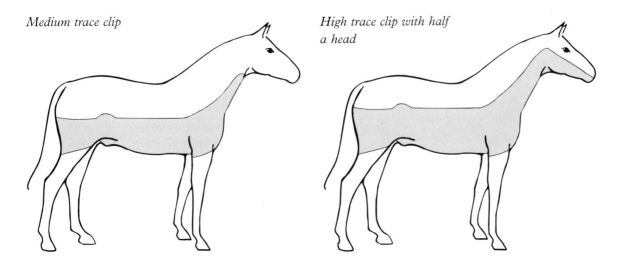

horse is very vulnerable and loses a lot of heat because of the thin covering of skin over the bone. If the horse's stable is not in a very sheltered spot or if you live in a very exposed part of the country, then a horse with a clipped head could suffer from a severely chapped and sore face, so leaving on some protection would be much kinder.

HIGH TRACE CLIP

The high trace clip should really only be used on a horse who is stabled and turned out for a few hours during the day. This clip is ideal for horses who are in steady work and required to do fast work sometimes, maybe entering competitions through the winter. There is still some coat left on the neck and back for warmth and protection while the sweaty areas are removed. Half the face or whole of the head can be clipped, depending on the conditions outlined in the description of the medium trace clip. The height of the clip could be about 5 inches (approx 12.5 cm) above the bottom of the saddle flap. Stable and New Zealand rugs are required and, again, extra rugs or blankets may be added in bad weather. This clip gives the illusion of making the legs appear longer on a short-legged horse.

CHASER CLIP

This clip is very similar to a blanket clip but the hair is not taken off the upper part of the neck, and the clip finishes just behind the ears. This helps to keep

Chaser clip

Blanket clip with half head left unclipped.

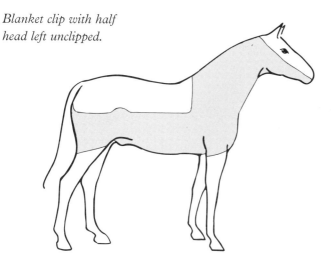

the level of a blanket clip. It was used widely on steeplechasers, hence the name. The chaser is used on stabled horses in medium to hard work such those competing or racing. Warm stable and New Zealand rugs are essential with extra rugs when weather conditions deteriorate. With such an extensive clip a close eye must be kept on the horse's condition; the same applies to the following clips, where most of the coat is removed.

BLANKET CLIP

This clip is so named because the winter coat that remains unclipped on the back, loins and rump resembles a blanket shape. The hair on the head, all the neck,

warmth in the muscles on the top of the neck. The coat is removed from the head, lower part of the neck, chest, belly and the upper part of the hind legs. The height of the clip is usually slightly above

Henry before…

chest, belly and upper part of the hind legs is removed. This clip should only be used on a stabled horse and it should be doing a great deal of medium to hard work and be in regular competition to warrant this clip. It still, however, provides some warmth and protection against rain on the back when out riding. Depending on the individual horse, an exercise rug will usually only be needed if the weather is very cold and/or wet and no hard/fast work is planned.

The height of the blanket clip can vary: a normal blanket clip is taken to the level of the bottom of the saddle flap, whilst a high blanket clip is above the level of the bottom of the saddle flap. It is usual to clip all the head, but half the face or all the head can be left on for some protection or if the horse is likely to feel the cold. If the horse is turned out during the day then a neck cover can be attached to the New Zealand rug when the weather is bad – the neck acts in a similar way to a radiator and quite a lot of heat is lost from this area. A stabled horse will also need extra rugs during cold spells. The blanket clip gives the impression of the back being much shorter so can improve the appearance of horses that are long-backed.

HUNTER CLIP

The hunter is another clip which is only necessary if the horse is in very hard work, and competing, racing or hunting regularly. Only a saddle area, legs, mane, tail and a small triangle above the tail are

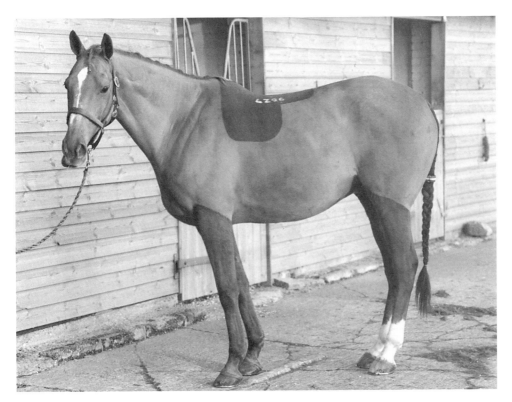

...and after his hunter clip.

Hunter clip

Full clip. Leave mane and tail and small 'vee' at top of tail.

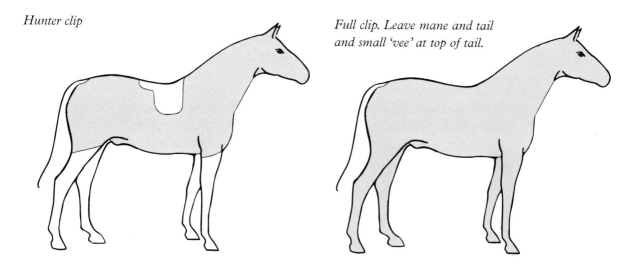

left unclipped. The pros and cons of leaving the saddle area on are discussed elsewhere. The horse should not lose condition with this clip providing it is kept in a good stable, is well fed and rugged appropriately. A horse can still be turned out for a few hours but extra blankets under the New Zealand are a must and a neck cover could be a neces-

The inverted 'vee' that is left at the top of the tail on a hunter or full clip.

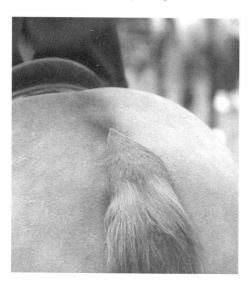

sity even in fairly mild weather. An exercise rug will be needed while out exercising.

FULL CLIP

With the full clip, all the coat is removed except the mane, forelock and tail. The legs and saddle area are clipped off. Sometimes the horse is given a full clip in the autumn and then for the second clip, another style of clip is given, possibly a hunter, blanket or even a trace. This will leave more insulation than the original full clip but the coat remaining is still much shorter. Stable bandages are needed to keep the legs warm to compensate for the removed hair. The full clip should only be used for horses in hard, fast competitions, racing and hunting. If the horse is turned out, then it should only be for short periods on fine, mild days and wearing plenty of warm rugs and a neck cover. An exercise rug should definitely be used while out exercising. Sometimes the full clip is given to show horses in spring so that the coat is short for early shows.

MARKING AND CLIPPING LINES

Horses and ponies are all different shapes and sizes and can definitely benefit from being clipped by someone with a good eye for lines and proportions. If you study the horse's shape and outline, the clipping lines usually present themselves – e.g. it is possible to 'see' the ones at the top of the forelegs and sometimes even the shape of the saddle patch.

Some horses have very poor conformation, such as a dipped back, a ewe

Clipping a straight line. Run the clippers along the line marked by the chalk. With experience you can achieve a neat straight line first go, but usually some tidying of the lines is necessary afterwards. This can be achieved by one of two ways, as shown in the following two photos.

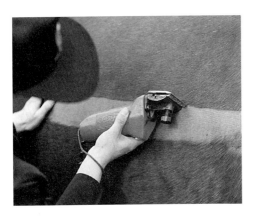

To neaten an edge, turn the clippers 'side on' so that the blades neatly cut along the jagged edge.

Another method of tidying a line is to place the blades along the line and lift away, allowing the hair to run through the teeth.

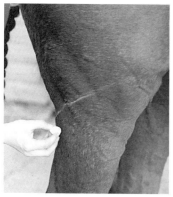

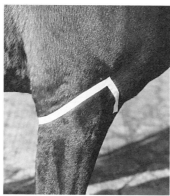

Lines can be marked off with chalk (above left) or tape (above right).

neck, a goose rump, or are too long or too short in the back. Others can be odd shapes due to poor nutrition or inappropriate exercise, which results in pot bellies, being very overweight or on the lean side. As well as sorting out the fitness and feeding side of things, the appearance of these horses can be improved greatly if clipped by someone experienced.

When you have finished the lines on a blanket or trace clip, the main body line will look straight when the horse is standing still but when the horse is working the line often appears to slope downwards. To prevent this, try taking half to one inch (12-25mm) off the line from the shoulder to the elbow. This will give the

illusion that the line is straight on movement, when in actual fact it is not. 'Tricks' like this come only with practice and a good teacher.

Lines are always difficult, even for the most experienced person and it can be very frustrating trying to get them level on either side of the horse. One side is inevitably higher than the other, so it's just as well that you cannot see both sides at once!

Practise clipping the edges and body lines on parts of the coat that you will be clipping off eventually, so it won't matter too much if you make a mistake.

There are some things that can be done, however, to help obtain a straight line.

○ Chalk or a coloured marker can be used on the coat to draw the lines.

○ A bar of damp saddle soap can also mark the lines on the horse.

○ Adhesive tape can be applied, but some horses do not like this, and it is not effective on a horse with a very thick coat or one who fidgets as the tape tends to crumple. Tape will only stick to a horse's coat if it is very clean.

○ Make sure that the horse is standing on a level piece of ground and is standing squarely on all four feet. Many horses and ponies like to rest a hind leg, especially if they have been standing for a long time. It can be very annoying if you think you have clipped a good straight line only to find that it is crooked when the horse stands up square again.

○ Sometimes a measuring stick (or something similar) can be employed to

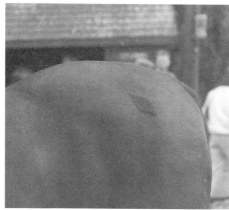

Clipping can be as individual as you wish to make it – sometimes an artistic streak comes out. This diamond was left on just for fun.

Putting tape along the horse's body. Make sure that the horse is clean otherwise the tape will not stick on.

A trace clip marked out along the body and legs.

get the level fairly equal on both sides, especially on a trace or blanket clip. For this to be successful, the horse or pony must again be standing on level ground, be absolutely still and not be afraid of the measuring stick and shy away from it.

❍ Never try to get a good edge to the lines with blunted blades, the end result is always untidy. To obtain a neat edge, change the blades to a sharp set and then finish the lines. Always clean and lubricate the blades before finalising the lines; this will guarantee a neat finish. Get all the hair going in the same direction by brushing down the rough edges along the line before clipping.

CLIPPING THE LEGS

There are arguments for and against removing the hair from the legs.

Personally, I feel that it depends on the circumstances, such as breed, work requirements and type of management.

Horses with clipped legs must always be stabled and in hard work. Leaving the hair on the legs can offer protection from the cold, wet and thorns. Possible cuts may not be so severe if the legs are shielded by the hair. On the other hand, if the legs are clipped, then any injuries can be seen straight away, and any thorns that have penetrated the skin can quickly be detected and removed before the leg becomes infected. The legs can also be dried off more quickly and more thoroughly if the hair is removed. Heat loss from clipped legs can be quite a problem, though, so it is necessary for the horse to wear stable bandages when not working.

Some say that the role of the feathers at the back of the fetlock joint is to act as a drain for sweat and mud, so that the heels are kept dry and therefore the possibility of cracked heels occurring is reduced. The 'drain' theory is questionable because horses and ponies kept out at grass during the summer can suffer from cracked heels too. If, however, you do hold with the drain theory but still want the legs clipped for easier drying, then a small tuft of feather can be left on at the back of the fetlock joint.

If the legs are very hairy and difficult to dry but you feel that clipping would be beneficial and are worried that some insulation and protection is needed, then the legs can be clipped with medium or coarse blades, which will leave some length of hair on the legs.

Native ponies or light horses that are crossed with the heavier draught types, may have extremely hairy legs with a lot of feather and this makes drying the legs and heels properly very difficult indeed, so if these types are stabled and are working quite hard through the winter then clipping the legs may be necessary. Remember, though, that if the native pony is registered and is to be shown in its breed society classes in the spring, then be aware that it has to be shown in its natural state, that is with full feather, and therefore clipping the legs will reduce the pony's chances of being placed or winning.

If you do decide to clip the legs, remember that they may be more prone to rubbing from protective boots, especially if you work on all-weather surfaces. Sometimes the material used for the surface can find its way into the space between the legs and the boots, causing chafing. This problem can be exacerbated if the legs are clipped because the bare skin has less protection. The insides of the boots must therefore be cleaned thoroughly after every use, and the legs carefully cleaned and dried.

Sometimes the surfaces of 'do-it-yourself' all-weather arenas can be quite harmful as the materials may originate from dubious sources. They may contain chemicals or substances that can burn the horse's legs, which is why cleaning the irritant from the legs every time the horse is worked is so important, and even more so if the legs are clipped. This is more evident with 'sand-topped' surfaces, so watch for any signs of inflammation or discomfort when the legs are brushed. With professionally made all-weather arenas this is rarely a problem.

Mud fever

When a horse or pony develops mud fever, sometimes the vet advises clipping the hair from the affected area so that treatment can be given more easily. When clipping, be careful not to use any blade washes or too much oil as this can drip onto the infected sore area, sting the horse and exacerbate the condition. If a barrier cream has been applied prior to the decision to clip, then this must be carefully removed and the area cleaned and dried thoroughly. The blades may not clip through the coat if there is a layer of grease which can clog the blades.

Mud fever can develop into a serious infection which can cause the horse to become very lame. If allowed to get out of control it can spread up the legs and affect the belly. If you have a horse who has been seriously affected by mud fever in previous years, then be very observant with the legs and be wary about clipping the stomach, especially if the horse is living out. Horses and ponies with white legs seem particularly prone to mud fever, so keep a closer eye on these.

Check your horse's legs every day for any signs of scabs or inflammation. If your horse is living out and any scabs are noticed, then take the horse out of the wet mud and into a dry stable. Soften the scabs by washing the legs in an antibacterial shampoo and water before you remove them. Only when the scabs have been removed should you clip the area, but be very careful because removing the scabs will leave the legs extremely sore and they may already be weeping and inflamed. There is a risk of being kicked whilst attempting this, so be very careful and take precautions, e.g. holding up a front leg if clipping the back legs; your vet may prescribe a mild sedative if the horse will not let you near his legs because of the pain. Treat the area with whatever your vet advises, which is usually an antibiotic ointment.

Prevention is always better than cure and good management can save a lot of misery, pain and distress. Keep the horse out of the mud as much as possible. Do not turn him out for hours on end just to stand in a boggy field. Ideally the horse's legs should be kept dry and clean at all times as the infecting agent cannot survive in a dry environment. This is not always practical as horses have to be ridden, but if your horse is susceptible then be even more particular with drying off after exercise.

If you have a manège then use this as a turn-out area so the horse can have free movement for a few hours but the legs are not soaked with wet mud which will precipitate an attack of mud fever. Turning the horse out is very important for his mental welfare. The horse should not be kept confined just because he has mud fever. Be sensible – turning out is fine if the problem is not very severe and he is receiving the correct care and treatment. Leave him out long enough to let off steam and bring him in straight away when he settles down. Apply a water-resistant barrier cream, such as zinc and castor oil, E45 cream, liquid paraffin or udder cream, if you have no other option but to turn your horse out in muddy conditions.

THE SADDLE PATCH

There are various schools of thought as to the usefulness of the saddle patch. The hair left on the back can act as a protection against the saddle rubbing during a long ride or a day's hunting, but this should not be used as a substitute for good management. Saddle sores occur for a variety of reasons, but a saddle patch may offer little, if any, protection against many of their causes. Most injuries are caused by pressure or friction from a saddle that has been poorly fitted; by bad, insensitive riding; or from a loose girth that results in the saddle being mobile on the horse's back. Saddle sores may also appear on weak backs which are not protected by a developed layer of muscle, especially on young, old, unfit or badly nourished horses.

Some horses are 'cold-backed', which means that they 'duck away' from the saddle when it is first put on or when the rider mounts. This also has a number of causes, such as damage to the back muscles or vertebrae, kidney infections, or a sudden drop in blood pressure when the

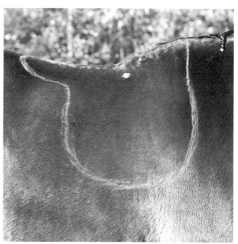

To mark out a saddle patch use the horse's own saddle and draw around it with chalk. Secure the saddle with the girth, and leave a margin of about 2ins (50mm) around the saddle so you have some coat to play with when you come to clip the outline.

Ready for clipping – the chalk-drawn outline of the saddle patch after the saddle has been removed.

girth is done up. Sometimes it may be a panic reaction – horses have long memories and they may associate the saddle with the pain of an old back injury, even though they have received professional treatment and are no longer in pain. Leaving on a saddle patch can help a little with this reaction whatever the cause, but the root of the problem should be investigated and dealt with if possible.

If the saddle area is removed, a very sensitive horse can sometimes behave similarly to a cold-backed horse because he objects to a cold saddle being put onto a bare back. This problem is lessened by using a good quality, comfortable numnah. Putting the saddle on a few minutes before mounting but not fully girthing up straight away, and only gradually tightening, may also help.

Sensitive horses can be allergic to the detergent that is used for washing the numnah, and with a numnah directly against a clipped back, the risk of irritation is greatly increased. A numnah made of natural fibres and washed in mild washing powder will be more comfortable for the horse and less likely to cause an allergic reaction.

The saddle patch could be clipped out with coarse clipper blades which would leave some length of coat but would make drying and cleaning this area much easier.

If you have a horse or pony who is sensitive to a cold girth after clipping and prone to girth galls, then a girth sleeve can be used to cover the girth. This is also best made from natural fibres.

Many people worry that if the saddle area is clipped off completely, then the short stumpy hairs may irritate the horse's skin because they are forced into the back by the saddle and the rider's weight. Although the theory is understandable, this does not happen, and most problems are usually caused by bad management. It is better, therefore, to leave the saddle area on but care must be taken to ensure that it is properly dried, as there is a possibility of chills. Always brush or wash the sweat out of the saddle patch, as the back can so easily become sore if the horse is ridden the next day without the sweat being removed. Sometimes the saddle area is removed initially when the horse is given a full clip, then the area is left on if, say, a hunter clip is done the second time around. The area is thinner so the horse does not sweat so much and it can be dried off much more easily.

When clipping the saddle area, the horse's own saddle must be used to make the shape. The numnah, stirrup irons and leathers should be removed but the girth can stay attached to the saddle so that it can be secured on the horse's back and not move about while the outline is being marked. The saddle should be slid back into its normal riding position on the

horse before securing. The saddle can be clipped around while on the horse's back if you prefer, but this is not recommended for novice clippers (horse or human).

Owners of racehorses or point-to-pointers often prefer to leave an 'egg' shape on the horse's back. This offers some protection over the withers and spine and is more in keeping with the outline of the racing saddle.

HOGGED MANES

The mane and forelock have evolved to provide the horse with warmth in winter and a heat shield in summer, and are ideal 'fly swatters'. The forelock, if left long, will protect the vulnerable eyes from flies.

The process of clipping off the mane and forelock is called 'hogging'. There may be many reasons for doing this. Ridden cobs (see page 10) – not Welsh Cobs – are shown with a hogged mane. You will also notice that polo ponies are always competed with hogged manes. The reason for this is that it prevents the

Clipping the the saddle patch.

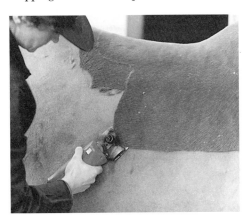

A finished saddle patch.

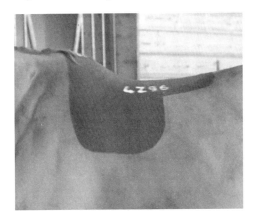

A hogged mane.

reins becoming tangled in the mane during the fast manoeuvres on the polo field.

Horses or ponies who object to having their very thick, unmanageable manes pulled or tidied, despite attempts at gentle handling or restraint, can have their manes hogged to improve their appearance. Also, horses with very thin weedy manes or manes that have been partially chewed or rubbed away, may be better off without any mane at all, providing they have quite a well-developed muscular neck. Horses with a very thin or ewe neck look odd with a hogged mane, so it's better to find other ways of improving their appearance.

The bulk of the mane can be removed with sharp scissors and then the rest can be taken off short with the electric clippers. If the mane is to be hogged in the winter and a clip that removes the neck hair is also done, then hogging is fairly easy. If a trace clip, for example, is required, then the 'join' between the

mane and coat can look very untidy unless very carefully clipped so that the end result is neat.

Most of the mane needs to be clipped from the withers upwards and the top from the poll downwards if the horse will allow. To make the shape of the crest, the clippers may have to be held vertically and brought down onto the top of the neck. This should be done on the near-side and the offside, but be very wary of the short, loose hairs flying off and hitting you in the eyes – they can feel like mini harpoons!

The forelock can also be started by removing the bulk with scissors so that the hair doesn't fall into the horse's eyes, and then finished with the clippers. If the eyes are particularly vulnerable and there is no really good reason for removing the forelock in summer, it is kinder to leave this on so that the horse is not plagued by flies.

Hand clippers or quiet cordless ones are useful for hogging the mane and forelock, especially if the horse doesn't like clipping. The mane should be clipped every ten to fourteen days to keep it short.

If you decide to re-grow the mane, be aware that it usually comes back twice as thick! Start training the mane by damping it down as soon as it is a few inches long. Be very persistent, then you won't have to put up with a wild and unruly mane later on.

Be careful with rugs over the withers where the mane has been removed. Without any mane for protection, there is a risk of the rug rubbing the withers, which may eventually result in fistulous withers if nothing is done to reduce friction. This also applies if just the base of the mane is trimmed for cosmetic purposes.

A hogged mane is not recommended for a horse or pony living out through the winter as the neck will be vulnerable to the wind and rain. In this instance, leave the mane as long as possible to help conserve heat.

Sweet itch

Sweet itch looks very unsightly and requires careful management to alleviate the irritation and distress it causes. Hogging can improve the overall look of horses with sweet itch as most of the mane on the lower half of the neck is usually lost or rubbed away. It also makes application of any treatment in the form of washes and lotions a lot easier.

Be careful when clipping the affected areas as the skin may be very sore, dry and ridged, so clipping could be very painful. Be extremely gentle, understanding and sympathetic if the horse shows any reluctance during clipping. As with mud fever, don't use excessive oil or blade washes.

If you know the horse is prone to sweet itch, then clipping can be done at the beginning of the spring before it becomes too bad. With good dietary management, treatments to prevent the biting of midges (*Culicoides pulicaris*) and keeping the horse or pony in during the times when the midges are most active (daybreak and nightfall), there is a good chance that the horse can grow a fairly normal mane and tail.

If, when buying a horse, you are offered an animal with a hogged mane in late autumn or winter, be wary. This may be an attempt to cover-up the fact that the horse is prone to sweet itch in the summer. Always ask the person selling the horse about the reason for hogging the mane.

Because the mane and forelock protect the horse from flies, some artificial protection must be employed if these natural accessories are removed. This can be in the form of a fly fringe which is put onto the headcollar. (Make sure that the headcollar is easily breakable if it gets caught on anything in the field.) Plenty of long-lasting fly repellent that the horse is not allergic to can also be used.

The clippers should never be used to clip the top of the tail.

·3·

LATE CLIPS AND SUMMER CLIPS

LATE CLIPPING

The general rule about when to do the last clip of the winter, is that it should not (in the Northern Hemisphere) be later than early February (some say late January), because the quality of the summer coat will be poor. If, however, the horse is still in hard, fast work then he may have to be clipped after these months so that he does not become overheated and distressed. The weather is usually slightly warmer by the end of February so he may begin to sweat up more quickly, and keeping the coat short will help him stay cooler longer.

The lines that show through the growth of a new summer coat can be avoided if the horse is clipped right out, which it should be if doing fast, competitive work. A glossy summer coat may not be important to people who use their horse solely for competition, but for those who do wish to enter their horses for show classes, there are many factors that go towards obtaining a coat in excellent condition. If the horse is fit and healthy, well groomed, on a balanced diet, free from parasites (both internal and external), warm and happy, then there is every chance that the coat will be restored to show condition. As well as feeding nutritionally balanced products that go a long way towards maintaining a healthy horse, there is also a multitude of supplements and shampoos available nowadays that are excellent coat conditioners.

Many show horses and ponies are given one full clip in spring so that the coat is short for early competitions. Take advantage of the warm, sunny spring days to turn the horse out for as long as possible. Sunlight is essential for the production of vitamin D, which will promote healthy skin, which ultimately influences a healthy coat.

SUMMER CLIPS

There are various pros and cons of clipping certain horses and ponies in summer. There are probably more people against this idea than for it, but it depends on circumstances, and the comfort of the horse must always come first.

Some countries, and especially the British Isles, have many native ponies who shed their winter coat but still retain quite a long thick, summer coat. There are also finer, more athletic horses that are crossed with heavier draught types and the offspring can 'inherit' the tendency to throw a dense summer coat. Horses sweat very easily on warm summer days and even if we ride in the evening to avoid the hot sun, a horse with a longer than average summer coat can become chilled if he sweats up and cannot be dried off properly in the cool night air.

Everyone knows that native horses and ponies such as the Welsh Cob, Fell, Highland and Shetland to name a few, have evolved over time to become adapted to the cold, wet and windy areas of the British Isles. Other European countries, especially the more northern ones, also have their hairy equivalents who likewise need to grow a dense, long coat for warmth and protection against the harsh conditions that are encountered in these areas.

Even the summers on exposed moors and mountains can be bitterly cold and wet, so protection is still important through these months. Just because we have taken the majority of these natives from their natural habitat and turned them into 'soft' pampered equines by providing stables, rugs, lush pasture and 'designer' feed products, it does not necessarily mean that their coat growth patterns, evolved over millions of years, will be cancelled out in around a century. In their natural environment, native ponies have little cause to sweat and

their whole evolution is geared to heat conservation and not heat loss. They are therefore more likely to overheat when worked or asked to undertake fast work in hot conditions than would an Arab horse which has evolved in desert climates. The versatility of the native type is well known. Increasingly they are being successfully competed and they make excellent children's mounts, so therefore they may need a little help to cope with all the faster work they are expected to do. Many considerate owners will withdraw their horses/ponies from competition in sweltering conditions, but this can be very frustrating if points need to be accumulated to qualify for a final, or if the classes are qualifiers for a bigger event, so we need to help them all we can while competing.

To discover how British native ponies with long coats fare in hotter climates I talked to Welsh Cob and pony breeders who have sold animals to very warm regions in Australia and the USA. The new owners observed that the ponies still grew a longer coat (but not a full winter coat) in the same months as the British winter, even though the temperature was very warm. It seems as if the ponies' 'inner clock' was still influencing the coat according to the British seasons. For some of the cobs to be able to work comfortably without overheating, it was necessary to clip them when they grew a longer coat. Exported ponies do eventually acclimatise to their new country; it usually takes around two/three years, although this varies with individuals.

Why else do horses need a summer coat of varying lengths? Wouldn't it be easier if they were completely bald? Some coat

Lynn Russell with her grey ridden show cobs. Ridden cobs are given a full clip in the summer and their manes and forelocks are clipped off (hogged). The grey on the left has not yet been clipped. Compare him to the very smart clipped grey on the right. Fine blades are used to give a close cut.

is actually still needed for protection against fly bites and injury, such as bites from other horses. It is better that a horse who attacks another goes away with a mouthful of hair instead of a mouthful of flesh! The coat also protects against the sun, and some horses have been known to suffer from sunstroke. More and more horses and ponies are becoming sensitive to sunlight and this is evident by the increase in skin complaints and irritations. This is occurring in humans also and is a direct result of the thinning of the protective ozone layer and the gradually increasing levels of harmful ultra-violet rays.

Horses who lack pigment, for example pink-skinned or part pink-skinned horses, albinos, and chestnuts with white socks and a blaze, are particularly vulnerable, and summer clips on these horses can leave them very exposed to sunburn. These horses lack the pigment 'melanin'

in their skin; this protects them from the harmful effects of sunlight. If you own a horse of this type, and after great deliberation you still feel that clipping would be beneficial, then be very careful with his management. Bring him in at the hottest part of the day or put on a summer sheet which will help protect a little against the heat of the sun. Look out for any signs of distress and inflamed skin on white areas. Your vet should be called immediately if sunstroke is suspected. Affected horses should be taken out of bright sunlight and put into a dim stable until the vet has seen the horse and instructed you on the best course of treatment.

Horses and ponies can become photosensitive if they eat plants such as St John's wort, clover and trifoliates in hay or pasture. This condition sensitises the pink skin to the rays in sunlight. If your horse is susceptible, it may be worth finding out more about which plants cause photosensitivity and checking your grazing area. You should eliminate the culprits, if they are present, before any summer clipping is attempted. Thiabendazole wormers can also cause sensitivity, when fed with clover and other trifoliates. Photosensitive horses are prone to sunburn even when unclipped, but clipping may increase the problem.

Summer showers and storms can be quite harmful at times, because the rain contains pollution and irritants. The number of horses who show allergic reactions after summer showers, for example being covered in lumps and/or losing coat, is on the increase, so horses that have summer clips will definitely need some sort of protection or may need to be stabled during rainy weather. Ready-made, light, durable, waterproof rugs are available today, or they can easily be made to order. It may be worth investing in one of these as a precautionary measure.

To protect against flies and midges is very difficult, and they can be a nightmare even to a horse who has a coat for protection. For a clipped horse the problem is increased enormously. Plenty of fly repellent may be the answer (provided the horse is not allergic to this), or being brought into a stable for the day and being turned out at night when the fly population is not so active. Feeding garlic also helps to discourage flies as the garlic is secreted through the pores in the skin and is thought to repel the flies. Tea tree aromatherapy oil is also a very good fly repellent as flies do not like its 'antiseptic'-like aroma.

Even during summer the nights can be quite cold, and if the horse has a lot of coat removed he may require a rug to keep him warm. If you have a horse or pony who grows a thickish summer coat, even though he has been stabled during the winter, and you don't want to summer clip, then you can reduce the thickness by not roughing off quickly and turning out for good too soon. If you keep him well fed and well rugged up in the spring, both in the stable and when turned out for a few hours, the chances are that the coat will not grow back so thickly as in previous years. The horse can be left out day and night when the weather is warm and settled, possibly with a light waterproof rug. Watch that the horse is not uncomfortable and does not overheat, however.

The length and quality of the coat also largely depends on the condition of the horse or pony. Horses and ponies in poor condition take longer to shed their winter coat in an attempt to conserve warmth, which compensates for their lack of insulating layers of fat. Also, if the horse is 'wormy' he will hang onto his coat longer, so work out an effective worming programme – every six to eight weeks.

Even horses that are warmblooded may need to be clipped in summer, so that they do not become overheated and distressed during fast work. An obvious example is the Thoroughbred flat racehorse. They usually have very short, fine coats, but trainers will always cater for an individual horse's needs and they may have horses in training with a thicker than average coat. In this instance, the horse will benefit from a clip so that training and the actual races are not so physically taxing. Because racehorses are stabled whilst in training, the risk of photosensitivity and exposure to rain-borne

Native breeds retain a longer, thicker coat in summer, and may need clipping if they are asked to do fast, competitive work when it is hot. However, do bear in mind that registered native breeds taking part in breed society classes must be shown in their natural state (i.e. untrimmed).

pollutants are minimal.

If showing, then a full clip is necessary, but otherwise choose a clip which entails removal of the minimum amount of coat to do the job. If clipped right out, don't leave a 'vee' at the top of the tail as you would for a full or hunter clip. Try to blend into the top of the tail so that the clip is not obviously noticeable. Again, the horse should be clean and can be bathed the day before, but use a very

mild, gentle shampoo so that the natural oils are not stripped from the coat. Shampoos made from natural products are ideal. Clipping is much the same as for winter but because the weather is warm, the clippers are likely to heat up fairly quickly. They should be cooled and lubricated very frequently. If the coat is extremely long, but some thickness would be deemed beneficial, then medium blades can be used.

.4.

CLIPPING DIFFICULT HORSES

REASONS WHY SOME HORSES ARE DIFFICULT TO CLIP

Before attempting any clipping, it may be an idea to gain an understanding of some of the reasons that can underlie the horse's objection to the clippers. Horses are naturally suspicious; this is one reason why the species has been able to survive for millions of years. Clipping the horse is very unnatural and can sometimes be viewed by them as life threatening, so it is always surprising that owners expect the horse or pony to accept this without making a fuss and go into 'panic mode' when the horse does react violently. All horses and ponies are individuals and you do not always know how the horse is going to react when faced with the prospect of being clipped. A horse's normal, everyday temperament and personality is nothing to go by, as the most placid of ponies can turn into a 'demon with four legs'! Even the smallest of ponies is much stronger than a man, so it is much better to outwit them by common sense than by having an enormous battle of strength in which we feeble humans will always be the loser.

The reasons why horses are bad to clip depend largely on temperament, past experience of clipping, handling and training. A horse for sale is often described as 'has been clipped'. This is misleading as you never know if the horse has accepted being clipped easily, what methods were used or what the horse went through to get the job done. It is always better to deal with what the horse or pony presents you with now, and not to go by any previous history, however favourable it may have been, as there are many factors, even ones unrelated to the clipping process, that can change the horse's attitude to being clipped. If you are unsure of the horse's response, then always treat him as a youngster being clipped for the first time, even if you know the horse is about fifteen years old. There are horses of this age, and older, who have never come into contact with clippers.

In some cases, it is difficult to know whether the horse or pony is genuinely scared or is just being awkward. Telling the difference sometimes only comes with experience and knowing the reasons behind his behaviour. Only by observing and learning why horses react in a particular situation will you know when to reassure and when to discipline, and some horses may need a mixture of both. Don't take it for granted that the horse is misbehaving just for the sake of it. Horses that do this are very few, but it is usually our first thought. Equines are naturally cautious but they are also inquisitive, playful, fun-loving animals and we can use this to our advantage. We must attempt to make clipping as pleasurable as possible, not a military exercise. Always give the horse the best possible chance of having a good, reassuring experience with regard to clipping and this is all any horse or pony can ask of you.

List of reasons why horses dislike clipping

❍ Young horses or horses who have **never been clipped before** are likely to become very nervous and frightened at the sound and feel of the clippers.

❍ Some horses who have been b**adly handled and frightened** will inevitably associate the noise and feel of the clippers with a very terrifying experience. Rough treatment can result in putting the horse or pony off being clipped for life and it will take much time and patience to regain his confidence. Some horses are never one hundred per cent again about being clipping after being handled this way. It is therefore very important that the horse a given a good

experience when he is clipped for the very first time. He will usually accept clipping again without any problems. Horses that have been ill-treated and abused may take a long time to trust anyone, so it is understandable that they may be apprehensive about being clipped.

❍ **The skin of a horse is very sensitive** to the touch; this is because it is a minefield of nerves. Evidence of its delicate sensitivity can be seen when the horse knows where a fly has landed, when he responds to the lightest pressure of the rider's leg aids, or when he reacts to the slightest changes in temperature; horses also feel pain much more than do humans. So imagine how intense the sensation of the clippers must feel, especially if used by a heavy-handed operator.

❍ **Certain horses can be very ticklish** when being groomed, especially warmblooded, thin-skinned horses such as Arabs and Thoroughbreds. Even to some of the cold-blooded, thicker-skinned native ponies and horses crossed with draught horses, the clippers must feel awful. Some do not like being groomed at all and show their annoyance in no uncertain terms. It is therefore not surprising that ticklish horses are a problem with the clippers. An experienced person, who is used to clipping these horses, should have the 'right touch', so clipping is not too difficult. Don't be tempted to put the clippers on the horse too lightly as this will almost certainly feel ticklish, and this is what you want to avoid. Put the clippers firmly onto the coat so more of an even pressure is felt.

❍ **Nervous horses** are also awkward

because they tend to shudder and tense up when the clippers are put on them. This makes a smooth, even clip difficult, and straight lines virtually impossible. These horses are a problem because they will inevitably be reluctant to stand still. Nervous horses need firm but sympathetic handling and constant reassuring, but they may not respond to this if they are too upset and on the verge of a 'panic state'. Some form of sedation may be the key to these horses or many hours of practice sessions, gradually familiarising the horse or pony to the sound and feel of the clippers. These horses are especially difficult because they tend to sweat up quickly so that clipping becomes impossible. Care must be taken not to let the horse reach this stage. Stop as soon as you see the first signs of sweating, such as the veins standing out, the coat feeling warm or damp patches appearing on the neck. Begin again when the horse has had time to cool down and become settled.

❍ It is unreasonable to expect young or highly strung horses or ponies to stand still for the amount of time that it can take to do a clip from start to finish, because their **concentration span is fairly short**. Even if they start off obligingly, their willingness to co-operate can soon diminish. Having an argument about this will only lead to problems. If clipping is done in 'blocks' of 10-15 minutes' duration (some horses may not even tolerate this length of time) with rest periods in between, then the horse should be more willing to stand still during the clipping sessions. Sometimes clipping may have to be done over a few days.

❍ Some older horses will start off behaving well but **become bored** quite quickly. Even the most patient of horses has its 'breaking point' with regard to standing still for hours on end. Bullying an otherwise amiable, well-mannered horse into standing still, may only serve to sour his attitude and turn any future attempts at clipping into a battle of wills. Again, give the horse a regular break.

❍ If a horse is **allowed to become cold** while being clipped, he will start to become restless and uneasy. Look for signs of the clipped coat standing up and looking 'starey'. Make sure that you have rugs nearby to throw over the back to prevent him becoming cold, especially if a lot of coat is being removed. If there is no other option, and you have to clip on a very cold day, try to keep the rug over the horse throughout clipping even if just a trace clip is needed. Fold the rug away from the area that is being clipped then put it back over. You will be surprised at how quickly the horse can get cold.

❍ Many horses do not actually mind the noise, but **hate the feel** of the clippers, whilst others object to the noise but can become settled and co-operative when the clippers are placed on them and they realise that they are not going to be harmed. You will soon find out which type your horse is!

❍ Sometimes horses **feel too well** in themselves, mainly due to overfeeding and lack of work. This causes many behavioural problems, not only with being clipped. It is advisable to evaluate the horse's diet in relation to his work requirements so that management problems as a result of overfeeding do not occur. If conditions allow, try to turn the horse out the day before clipping, either in the field or school so that he can 'let off steam'! But make sure he is cleaned thoroughly afterwards. If this is not possible, then take the horse out for a long ride the day before, never on the day of the clipping as he may not be dried off properly for clipping. The horse should be in a better frame of mind with regard to standing still and not fidgeting. A few horses may be more difficult to clip the second or third time around because they are usually much fitter and may be feeling very well in themselves.

○ If a permanently field-kept horse or pony is brought into a stable to have a small amount of coat removed, he may become worried and anxious because of **not being familiar with an enclosed environment**. Some horses suffer from claustrophobia and these horses are much happier if clipped outside in the yard if conditions are suitable. If you do have the use of stable facilities, bring the horse in each day so that he can get used to being inside the stable before clipping.

○ If a horse has to be taken to a **strange yard** to be clipped, he will inevitably be apprehensive and suspicious and may play up, even though if he is usually good to be clipped at home. Again, constant reassurance works wonders, and give him time to look around and settle.

○ Sometimes a horse may be **feeling 'under the weather'** although no obvious signs of illness are observed. He may not be feeling up to being 'pulled about' for the next few hours and objects to this strongly. Bear this in mind if the horse is usually very good to clip and no other reason for his behaviour has been found.

○ Some horses are **allergic** to blade washes and clipping oils, which can burn the skin and cause quite a lot of irritation and pain. Occasionally lumps and raised areas are evident so you know for sure that this is what the horse or pony is objecting to, but not always. Wash the oil off immediately if a reaction is observed; use cold water to reduce any inflammation. Bald patches can appear on the horse the next day or even a few days later, even if no obvious signs are seen at the time. Test a small area of coat a few days before you clip to make sure that the horse is not allergic to the products you have bought. By doing this, the horse will not associate any painful reaction with clipping.

○ Animals are far more **sensitive to electricity and electrical fields** than are humans. They can react to the electromagnetic field that electrical equipment emits. Make sure also that the clippers are not actually giving the horse minute electric shocks and have the clippers tested by a qualified electrician if this is suspected. Be very wary, too, of atmospheric conditions, especially thunderstorms. Humans often feel very tense and uneasy in thundery weather and it is the same for horses. This is because the atmosphere of the earth is being depleted of negative ions. Observe how the balance is restored if the thunderstorm is accompanied by lightning which appears to 'clear the air'. A full moon can have a similar effect for the same reasons. So avoid clipping if stormy, thundery weather is forecast.

❍ **If a horse is very dirty**, greasy or sweaty, the clipper blades will clog and 'snag' the skin, which can result in him becoming restless, occasionally bad-tempered and aggressive. Obvious signs of this are a slow sound to the motor, raised areas on the skin, blades not cutting properly and tufts being left on the horse; The clippers will strain on a dirty horse, this will make the blades heat up very quickly and can ultimately burn the horse if nothing is done to prevent this.

❍ **Blunt blades** are another common cause of horses objecting because they will catch and pull the coat instead of cutting through it cleanly, causing pain and discomfort. Again this will result in raised areas on the skin and an untidy clip. Never use damaged blades or blades that have teeth missing; these should be discarded immediately.

❍ **If the horse has a thick coat**, check that there are no hidden cuts, scabs, warts or growths that the blades will 'nick' when cutting through it. Some horses have understandably become quite difficult and nasty because of this happening on a previous occasion and they will no doubt associate clipping with the pain and discomfort they have experienced before. Never clip a horse with rain scald; see that this is treated and the horse's coat is clear from scabs before clipping.

❍ Horses are handled, mounted, led etc. on the nearside and as a result the horse will always be happier about this side being clipped and can become very **uneasy when the offside is attempted**. Make a point of handling the horse on both sides, so that when the time comes to clip, there will not be any problem with clipping the offside.

❍ Sometimes horses can be a problem because they are **very bad-mannered** and have been allowed to get away with misbehaving. They ignore requests to stand still, can be very 'pushy', barge the handler around the stable and crush the person clipping against walls and stable doors. Have with you a baton or stick (not too pointed) about 18-20 inches (approx 45-50 cm) long. When the horse does misbehave, the baton can be jammed against the wall and the horse's rib cage or shoulder, so he is unable to swing around enough to crush you. A couple of digs in the ribs like this will soon stop this behaviour. Basic manners, discipline and respect for the handler must be established before any clipping is attempted.

A thick winter coat can hide cuts, scars, and growths such as this, which can easily be cut by the blades.

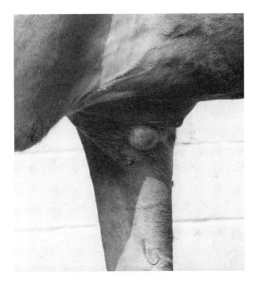

❍ One of the worst types of horse to clip is the one who is genuinely afraid of the clippers but who has a lot of character and is **bold enough to retaliate** against the handler and the person doing the clipping. Such a horse needs a very experienced person to clip him. Attempting to use excessive discipline will not be effective in overcoming his fear of the clippers. What he needs is a firm but patient handler who will not put up with any nonsense.

❍ Horses are **creatures of habit**; they get used to a particular routine, and more so if they are stabled. They can play up if clipping is tried when they know that this is the time they are usually doing something else, especially if it is an activity they enjoy, such as feeding, being turned out or their rest time in the afternoon. If this is suspected, then clip at a time when they would not mind missing that particular activity, such as schooling!

❍ Some horses that are '**cold-backed**' may be a problem if a full or hunter clip is planned, as they will object to the feel and pressure of the clippers when they are put onto the back. The cause of this should be investigated and treated if possible, but there are some horses for whom the cause has eluded investigation and remains a mystery. These horses must be handled very carefully and gently when clipping over the back. Sometimes, after the horse has been seen by a back specialist, the back may be a little tender for a few days. Try not to clip for at least five days, sometimes longer, depending on the problem and amount of work that has been done. This will prevent any discomfort being associated with the clippers.

❍ Certain horses are capable of deliberately **throwing themselves onto the stable floor** to avoid of being clipped. Some horses have learnt this evasion already in other situations, such as when being ridden or driven, and have adapted this behaviour to clipping. In such cases, the horse may still be clipped while he is down on the floor but this must **only** be attempted by an expert and is certainly not for the beginner. A horse has to raise his head and neck to enable him to stand. If you have an experienced and confident handler (which you should for a difficult horse), then he or she will be able to lean on the horse's head and neck (without so much force as to hurt the horse). This will keep the horse's head down so that clipping may be continued, provided the horse is not panicking. The person clipping should stand behind the horse, away from potentially flying legs. Make sure, too, that all the leads are out of the way. If clipping is continued while the horse is down, then the action of the horse throwing himself to the floor is not having the desired effect, which is to prevent to clipping being done. Usually, but not always, depending on the management of this behaviour if it occurs in other circumstances, the horse will probably discontinue this evasive reaction, sometimes for something more effective! If the horse is known to do this, then plenty of bedding should be put down and any stable fittings must be removed or covered. Also protective bandages and boots must be put on – even if you feel he deserves any injuries by deliberately throwing himself down!

❍ If you are going to clip late or have a summer clip on a mare, be very careful as the temperament and tolerance level of some mares can change dramatically whilst **in season**. A few mares become aggressive and can be quite dangerous, even under normal circumstances, so this danger is increased during clipping. Be very vigilant for any signs of irritability or temper. The mare may have to be sedated; otherwise an hormonal balancer may produce dramatic effects. Consult your vet about this; appropriate treatment could make life a whole lot easier!

❍ Horse and ponies that are turned out may have sustained a **recent injury** such as a bite or kick from another horse. The injury is not always visible if the horse has been wearing a New Zealand rug and still has all of its winter coat, but there may be a bruise forming deep within the tissue. During clipping, the pressure of the clipping machine is enough to cause the horse discomfort, so keep this possibility in mind if a usually well-behaved horse starts to flinch and no other reason can be found for the reaction.

❍ Even if a horse is been good about having his body clipped, there are few who will not show some **reluctance when it comes to having the head done**. There are many reasons for this:

❑ A horse's **sense of hearing** is far more sensitive than ours; they can detect sounds of a much higher frequency than we can, and they can hear very quiet sounds that we humans are unable to hear. Because of this fact, the noise of the clippers is extremely loud to the horse. It is not at all surprising that horses become very frightened when the clippers are nowhere near them, let alone in the vicinity of the head.

❑ Some horses are very **head-shy** through being physically abused. As a result, they may be very reluctant to allow anything at all near their head. They may even be apprehensive about brushes, headcollars and their bridle, so a noisy clipping machine is totally out of the question.

❑ Many horses may have a mild **ear infection or parasites** that have not been detected. These horses will not be happy about having their ears or head clipped as this could further irritate the problem. The root cause should be treated before any clipping is attempted. Never clip inside the ears, because the hair is a natural protection against flies and debris getting deep inside. At the very least this would cause an irritation and may precipitate a possible infection. For stabled horses there is always the risk of bedding going inside an unprotected ear when the horse lies down.

❑ Horses can also object if they have **sinus problems, toothache, sharp teeth edges** which scrape the tongue or inside the cheek, **throat infections** or **nerve disorders**. Like ours, horses' teeth can fracture if they bite on something hard; this can make the underlying nerves very sensitive. There are usually other indications of this such as evasions and head-shaking when being ridden, or reluctance to take the bit in the mouth when being tacked up. If your horse is exhibiting any of these, be aware that there may be potential problems with clipping the head. Again try to find out the reasons as correct treatment can make life a whole lot easier for yourself and less miserable for your horse.

❑ If there are problems with clipping the head, and again no cause seems obvious, be aware that the pressure of a horse's bridle, headcollar or cavesson may have caused damage and **bruising** to the underlying tissue although no signs of rubbing may be evident. Pressure from tack can easily occur. As little as a few ounces of continuous pressure per square inch on a horse's body will result in the blood supply to the surface area being reduced. If even more pressure is applied, such as 1-2 pounds (0.5-1kg), then after a fairly short period of time the skin and the tissue below it are starved of their blood supply and they begin to die. This is also how pressure sores under the saddle and girth are caused.

❏ **The ears of some horses are naturally sensitive**, even if there are not any identified reasons for this. If this is the case, the ears should be handled very gently. The ears could be left unclipped if they are not particularly hairy and the outside edges can be just trimmed with scissors to give a neat finish. The ears should never be twitched but some horses do respond if the ear opposite to the one being clipped is gently massaged.

❏ The head is usually last to be done, and by this time the **clippers could have become hot** and are burning the horse. The horse is much more sensitive on the head because there is a very thin layer of skin over the bone at the front of the face, so the heat and vibration of the clippers is always more intense, which at the very least is uncomfortable (and I would imagine very painful). Clean and cool the clippers before starting the head.

❏ If the head is left until last, then the horse is probably **fed up** and would rather be doing something else by now. Try doing the head when the horse has had enough time to become accustomed

to the clippers and will accept them on the face, and then return to the easier places on the body.

❏ It is surprising how many people clean the horse's body extremely well, but then leave the **head very dirty**. As the head is a notoriously difficult part of the horse to clip, this area should be as clean as the rest of the horse, if not cleaner. If left dirty, the blades will become clogged with hair and grease and become quite hot fairly quickly. If this is allowed to happen, the horse should take no blame for showing his annoyance. Pay as much attention to cleaning the head as you would the body of the horse. It is much easier to prevent problems occurring than to deal with them after the damage has been done.

SOME HINTS ON CLIPPING DIFFICULT HORSES

○ Make sure that the assistant is very experienced and knows what is required with regard to the handling of the horse; this will reduce the risk of injury. Ensure that he/she keeps the horse's head turned slightly towards the side being clipped. If the horse moves, then the quarters will swing away from the person doing the clipping.

○ It is better for the horse to be held than tied to a solid object where he may panic and pull back. If you have to tie up the horse for any reason, make sure you have used a quick-release knot just in case of emergencies. Don't ever tie directly to a ring or bar; use a piece of cord that will break or can be easily cut.

Part of a rubber inner tube can be used as this 'gives' with the horse and goes back into shape when the 'tantrum' is over.

○ Ensure that the horse is very clean and dry. This applies to any horse, but even more so for problem horses as this should reduce any reasons or excuses they have for misbehaving.

○ Check that the tension of the blades is correct. The wrong tension will make even the most placid horse become objectionable as this will pull the horse's coat and be very painful.

○ Don't allow the blades to become either clogged or hot. Again this will upset the horse.

○ Clip at a time when the yard is quiet and there is nothing happening that will upset the horse, such as other horses constantly moving in and out of the yard or adverse weather conditions like gale-force winds, for example.

○ If it is the first time that the horse has encountered clippers, or if he has been frightened previously or you have a nervous horse, then is a good idea to let him see another horse being clipped. It should be one who is very used to clipping and will not show any signs of becoming distressed or tense, however seemingly insignificant. This will help the horse or pony to get used to the sound of the clippers and he will know by the behaviour of the other horse that there is nothing to be alarmed about.

○ It may be a good idea to record the noise of the clippers onto a tape and play this to the horse while you are grooming,

Small, quiet, cordless clippers are excellent for getting a young horse used to the noise and sensation of the clippers. They are also useful for clipping areas that the horse is reluctant to allow a bigger, noisier machine near, such as the head.

then the sound at least will be familiar when you do actually clip the horse.

❍ Horses are naturally suspicious, but they can also be very inquisitive creatures and love to examine strange, unfamiliar objects, and the clippers are no exception. While the clippers are switched off, let him sniff the clippers and be allowed to touch them. Reassure him all the time and this will go a long way in helping to allay his suspicion.

❍ Before you actually turn the clippers on, run them over the horse's body for as long as it takes for the horse to accept them, and remember to talk to him all the time.

❍ Stand well away from the horse when you turn the clippers on just in case he 'spooks' or strikes out with a front foot. Make sure the handler is stood to the side and not in front of the horse for the same reason.

❍ Before the clippers are put on the horse, let him feel the vibration by putting your hand on the horse's shoulder and the clippers on your forearm. Gradually move the clippers down to the back of your hand and eventually onto the horse's shoulder. Again do not rush this stage as it can be the 'make or break' of how clipping will proceed. Move very slowly and quietly so as not to startle the horse. The clippers should be run all over the horse's body (or at least the parts you will be clipping) while switched on and keep doing this until the horse becomes accustomed to this before attempting to cut the coat. The sound of the clippers changes when the blades start to cut the

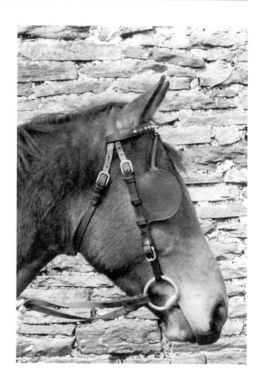

A driving bridle can be helpful in restricting the vision of horses who do not like seeing the clippers approaching. Be careful not place the clippers suddenly on the horse and take him by surprise.

coat; some horses may become startled at this so be very vigilant, just in case.

❍ Some method of distraction is also useful. Horses love music, so it may be an idea to have a radio playing nearby. This should not be too loud as the horse should still be able to hear your voice reassuring him. Singing or humming can give the horse something else to think about (especially if your voice is awful)! Only one person should be talking to the horse at any one time, either the handler or the person doing the clipping. By doing this, he does not become confused and his attention is on one person.

'Titbits' will also go a long way in distracting the horse or pony from the clippers, but should really be used as a reward for the desired behaviour, not as a 'bribe'. Some horses can misbehave deliberately because they know they will be given something nice to eat to make them behave. Most horses have a favourite 'itchy spot'. Find out where this is, and if appropriately situated, scratch this area; this can help to take the horse's mind off the clippers.

○ Horses sometimes panic when they see the clippers approaching but will soon settle when the blades are actually on them and they realise they are not going to be hurt. The difficulty is getting the clippers near them. A way of getting around this is by using a method of restricting the horse's vision such as a driving bridle or racing blinkers. These will also be useful in protecting the eyes if the horse starts to throw his head around. This can be a help to the person clipping because if the horse tries to kick, his aim should not be so accurate as his vision is restricted. A few horses, however, seem to know intuitively where you are standing, even if blinkered, so don't ever be lulled into a false sense of security and think he cannot get you! It is not recommended that you use a blindfold on the horse as this may make him panic even more. Don't ever use blinkers as a first resort, wait until you know how the horse is going to react. Shielding his view with your hand behind his eye is also helpful for these horses. Certain horses are happier if they can see where the clippers are, and are not suddenly taken by surprise when the clippers are placed on them.

○ If the horse is afraid only of the sound of the clipping machine, then cotton wool can be used in the horse's ears to muffle the sound. Care must be taken not to make the cotton wool pieces so small that they could get lost in the ears if pushed in too far. Make sure they are removed after you have finished clipping. Horses with sensitive ears may not let you plug the ears so another method of reducing the sound has to be found. Using very quiet cordless clippers may be the answer.

○ In the weeks running up to having the horse clipped, try using small, quiet cordless clippers so that the horse becomes accustomed to this level of noise, then the bigger, more powerful clipping machine won't seem so terrifying.

○ If you have an electric groomer, then this is useful for getting the horse used to the sound of something mechanical close to him and a different sensation on the skin.

○ If the horse does not like the electric clippers around his head, try using a set of manual hand clippers. This can be hard work and takes a fair amount of time so the benefit of removing the hair from the horse's head must be well worth the effort. Unless a horse is extremely fussy and intolerant he will usually let you use these on the head, providing the blades are sharp and cutting cleanly. The blades of the manual hand clippers should still be cleaned and lubricated frequently to remove the hair although they will never require cooling. These blades are removable and will occasionally require regrinding.

○ **Never lose your temper** when the horse or pony is misbehaving and you seem to be getting nowhere. If you do lose your cool, go away and give yourself (and the horse) time to calm down. Have a think about why the situation seems to be getting out of hand. Sometimes you may have to retreat gracefully to win overall eventually. You may have to clip the horse over a number of days if he is nervous or his tolerance level is very low. Don't let the horse think he has got away with anything, however. If you can, go over a place that he has already let you do, and praise him for letting you clip this area as if this is where you wanted to clip all along. Never leave the horse thinking he has won; this will only make things worse when you try again. Always leave the horse on friendly terms. Patience and perseverance are the only answer; things *will* become easier eventually.

Kickers

Kicking and bucking etc. have always been a horse's natural defence against attacks from predators. This is one reason why the species has survived for millions of years. Just because the horse has become 'domesticated', which means handled by humans and undergone training, probably since foalhood, does not necessarily mean that these defensive reactions will not be exhibited under extreme circumstances, and clipping can be classed as one of the extremes. Horses can view the clippers as an attack, so it is only natural that they will want to protect themselves. The horse's natural instinct for self-preservation can ultimately take over in this situation. Logical reasoning

does not enter into it nor that it is the clipping machine making all the noise and may of hurt them, not the person holding it. For your own safety, all horses should be viewed as potential kickers and it is too dangerous to think otherwise. Even the old placid schoolmaster, never known to have kicked anybody, can have his moments if 'nicked' by the clipping blades or startled by something unexpected happening in the yard. Make sure, then, that you maintain a healthy respect for the horse, but do not regard him with mortal fear, because the horse will pick this up and use it to his own advantage.

❑ Beware of 'cow kickers', horses that can kick sideways and/or forwards. These horses can catch you in the face while clipping the belly or near the sheath. One such horse caught me in the back when I was clipping the chest! Only the battery pack saved me from serious injury.

❑ Be very wary of clipping the legs of any horse, even more so if he has been known to kick before. Think to yourself – is the benefit to the horse going to be worth the risk? Usually the answer is no.

❑ Sedation (see next chapter) may be used for such horses. This will serve to take the 'edge' off any inclination to kick, and if he does try, his coordination will be affected so his aim won't be so accurate. If he does happen to 'connect' when he kicks, then the force of the impact should not be so great.

❑ Some sort of protective clothing can be used. A good-quality pair of chaps can be very effective in deadening the impact of a kick, and road studs will do less damage. Cricket leg protectors and headgear have been known to be used on occasions (not by me personally, but my horse's travelling boots have come in handy more than once). Wear a hard hat if you feel one is necessary.

❏ Never crouch down to clip as you won't be able to get out of the way quickly enough if the horse kicks out.

❏ Ask an assistant to hold up a front leg. The theory of this is that a horse cannot kick if standing on three legs. With most horses this is very true but there are always clever horses that will prove an exception to the rule, so still be very vigilant. Ask the assistant holding up the front leg to warn you if the horse is about to put his leg down so you can get out of the way.

❏ If it is convenient, the horse can be left unshod if it is a known kicker. This will lessen the damage done by shoes, road studs or stud nails.

❏ If leaving the horse unshod is not practical then sacking can be put over the horse's hooves and tied or bandaged around the fetlock, provided the horse does not object to this. This again will help to deaden the impact if the hoof connects with the leg (or some other part of your anatomy)!

❏ Give the horse an Irish clip. This is much safer for the person clipping because the back end of the horse, which is the 'danger zone', is left untouched with this clip.

❏ If the horse does not mind being enclosed and is not likely to charge off around the stable, try a 'wall' of straw or hay bales between yourself and the horse. This is usually stacked two or three high, depending on the height of the horse and the size of the bales. This offers good protection if the horse does kick. The person

doing the clipping has to be very supple and agile to be able to clip from some of the angles that this method will present, but it is very safe. Not to be recommended for claustrophobic horses, though!

HOW TO AVOID CLIPPING

If clipping is too daunting even to consider, then there are certain management regimes that can be adopted so that the length of coat is kept fairly short, and therefore manageable, with regards to excess sweating and drying off after exercise.

Look for very early signs of a change in coat colour and texture around September, and as soon as this is observed bring the horse into a suitable stable and keep him rugged (add warmer rugs as the weather becomes colder – he may be too hot with a thick heavy rug immediately). Of course, correct feeding requirements must be adhered to, but do make sure that he has plenty of exercise and turn-out time in relation to his feed, which will prevent him becoming too 'fresh'.

Just as the lengthening of the days determines the start of mares coming into season in the spring, the amount of sunlight available partly controls the growing and shedding of the winter coat. If we wish to keep the coat short, we can use this to our advantage by leaving the light on in the stable later in the evening and turning it on earlier in the mornings when it is still dark. By doing this, the horse's body-clock can be fooled into thinking that the longer days mean it must still be summer. Try to allow the horse as much natural sunlight as possible because it is more natural and beneficial to a horse's well-being than artificial light.

If the horse is to live out during the winter, and bringing him in early and keeping him pampered is totally impractical, then there is another option. Let the horse grow his full winter coat and only exercise at a quiet, gentle pace to reduce the likelihood of him sweating up. This can be difficult with a horse who sweats up easily, is very overweight or becomes very excitable when ridden. If the horse has a full coat, then it is very unfair to him to compete or do any fast work through the winter, but you can always make up for this by having him fitter for competition during the summer. Make sure, however, that the amount of work you give him is adequate for him to be mentally and physically healthy.

.5.

SAFE RESTRAINTS AND COMPLEMENTARY THERAPIES

Even if reasons for the horse objecting have been identified and understood and all of the suggestions in Chapter 4 have been tried, sometimes a method of restraint is necessary. Restraint is totally alien to the horse yet everyday equine activities such as leading and riding are forms of restraint. Restraint should be a last resort and the type and level of restraint should be appropriate to the temperament of the horse and the amount of resistance that the horse is showing.

Restraint should only be attempted by someone who is very experienced at using restraint methods. If horses are badly restrained, then this will only serve to increase their fear of being clipped. A few horses react violently to any sort of restraint, especially if it causes physical pain and it can sometimes exacerbate the behaviour we are trying to prevent. It is the horse's natural instinct to get away from pain, and if a bad reaction occurs people, as well as the horse, can be badly injured. Even with restraint the horse is always capable of fighting us and winning.

It is always the horse's choice whether he gives into being restrained or not. We humans always think we have 'won' when the horse is behaving but this is never the case. Usually, once a sensitive area of the horse has been clipped, and the horse realises that he has not been hurt in any way, then he will very likely allow you to clip that area again without any form of restraint. The only two reasons why a horse should be restrained are to prevent injury to handler or the person clipping, and for the horse to realise that there is no real danger if the clippers are in a 'taboo' area.

Restraint should never be used as a punishment.

The twitch

The most common type of restraint is the twitch, which is applied to the muzzle. Some horses who have had the twitch badly applied before, remember how uncomfortable or painful it was and resist it being putting on again. The twitch can be quite effective as the muzzle is thought to be an acupressure/ acupuncture point. The pressure releases endorphins, which are thought to be natural tranquillisers and painkillers, so the horse is actually being sedated by endorphins from its own body.

Don't start clipping as soon as the twitch is applied. Wait a few minutes so that there are enough endorphins released into the bloodstream to have the desired effect of calming the horse. Look out for the horse becoming sleepy, the breathing rate changing, and the veins standing out. Stand well to the side of the horse, just in case he strikes out with a front foot. After a while, the amount of endorphins being released slows down, so every five minutes release the twitch slightly and then tighten it again to stimulate the flow of endorphins.

The twitch should never be left on for more than fifteen minutes, and the cord must never be thin as this will cut into the horse's nose. The cord can be made out of plaited baling twine (not a single strand), a length of rope that again is not too thin, or a thick leather thong. The cord is looped through a hole which has been drilled at the top of a wooden rod about 13 inches (approx 33cm) long which is sometimes grooved for a better grip. The loop is put around the flesh of the upper lip and the wooden part twisted until the lip is tightly held by the cord.

When you put the twitch on, make sure that the outer edges of the lips are turned under as the sensitive skin inside the lip can be easily damaged. Be very careful when 'twitching' as some horses react badly as soon as it is applied but a few horses seem to become 'dopey' at first yet have a tendency to 'explode'

A humane twitch correctly applied – it is not cutting into the horse or pinching the sensitive lips.

without any warning – so be on the look out for this even if the horse initially appears calm.

There is another type of twitch which is made of metal and has a similar action to a nutcracker. This is supposed to be more humane and kinder than the 'loop'-type twitch but still must be used with care.

When the twitch is applied, the circulation of blood is cut off. To help restore the circulation, the muzzle should be gently rubbed when the twitch is removed. The ears should never be twitched. Not only is this ineffective, it is also very cruel as there are many small muscles, cartilages and nerves around the base of the ears that enable them to be moved in a variety of directions to distinguish sound location. These delicate structures can be severely damaged without much force with the horse ending up looking like a 'lop-eared rabbit'.

It is important to twitch a horse correctly, without causing him further pain and distress. This horse is obviously not worried about being twitched and looks quite happy.

The twitch can be used on relatively quiet horses that are extremely ticklish and won't let sensitive areas be clipped. It will help to take their mind off the clippers, but it must never ever be used as a punishment. Remember that there is a very good reason for their behaviour.

A 'neck twitch' can also be applied. This involves pinching a fold of skin at the base of the neck, just in front of the shoulder, with your hand and 'tweaking'. This area could also be an acupressure/acupuncture point as the horse responds to this method in a similar fashion to the muzzle twitch.

Sedation

Another type of 'restraint' is the use of sedation. This is especially helpful if the

horse is very aggressive or nervous to the point of absolute panic. Tranquillising will help to reduce distress and anxiety in a nervous horse and take the edge off an aggressive horse's temper. This is certainly a much safer option for all concerned. Your vet should always be consulted and will be able to give you professional advice about the pros and cons of sedation. It may be an option for horses who react violently to any other form of restraint. Modern tranquillisers are more effective than they used to be – previously they could cause the horse to sweat, wear off quickly and the horse could react violently and mostly without any warning. They only lasted about 15 to 20 minutes and repeated doses were needed, which was very inconvenient. Newer drugs last for about an hour, depending on the dosage, and drug-induced sweating is no longer a problem.

Tranquillisers can be given either by injection or by mouth. If the horse hates injections and is likely to become upset even before clipping has began, then sedation is best given by mouth. The effectiveness of the drug depends on many factors such as the temperament of the horse, whether he is nervous or aggressive, the type of drug that is administered, the dosage, individual susceptibility of the horse to the drug. A few horses can be worse with sedation, especially if they have a bad allergic reaction; this is rare, but it can happen. Some horses become distressed as the drug starts to take effect as they are frightened by the feeling of becoming sleepy and being in a 'twilight zone' and they try to fight against this.

If you already know that the horse or

pony needs sedation, then your vet can be contacted well in advance and all the necessary arrangements can be made. Never try to sedate the horse when he has already been frightened and is sweating badly. If he has become distressed, and you feel sedation may be of use, then make plans to clip with the help of sedation on another occasion.

When the horse has been sedated, wait a little while for the drug to take full effect. Signs of the horse becoming 'sleepy' are: a glazed, 'far away' look in the horse's eyes, lowering of the head, the bottom lip drooping, and in stallions, colts and geldings the penis is visible. As soon as the horse is sleepy, clip the difficult bits first so that the easier places are left if the tranquilliser starts to wear off. The horse often becomes unsteady on his feet and may be very 'wobbly'. Bear this in mind if you have to move the horse around the stable or pick up his front leg to clip underneath the forearm. Make sure there is plenty of bedding down, just in case the horse is over-sedated and falls over. The horse should also have boots and bandages put on for the same reason.

Using two ties

If you are faced with a horse who barges and won't stand still, two leading reins can be used to keep the front of the horse central, providing you have two rings or ties either side of the horse. The leading reins are attached to each side of the headcollar and are then tied or preferably put through the rings and held by two assistants. Beware of the horse swinging around and squashing anyone.

Using a bridle or headcollar

To have better control of the horse during clipping, it may be an idea to use a bridle rather than a headcollar. If using a bridle for extra control, the leading rein can be put though the bit rings to give a 'curb' effect. The method must be used carefully as a lot of damage can be done to the mouth, especially on young horses. Putting a leading rein or chain through the horse's mouth is not recommended as this can cause a great deal of pain and damage.

A Chiffney bit (see photo below) is useful for holding strong horses.

If a headcollar is used the leading rein can also be put through the bottom rings either side of the headcollar and this will again give a curb effect. For a very strong horse, the leading rein may be wrapped over the nose and then put through the

A Chiffney bit is useful for controlling rebellious, strong horses – especially those who like to rear.

A lead rein looped under the chin, giving a 'curb' effect, can help control strong, unruly horses and ponies.

rings, which puts pressure on the sensitive part of the nose. A lunge cavesson can also offer an effective means of control for a horse who lacks manners.

Working without restraints

Some horses object to restraint of any kind. They can become even more frightened because they feel 'trapped'. These horses are best kept 'on the move' and allowed to walk around you in circles, with you holding the leading rein in one hand and attempting to get the clippers on his shoulder with the other hand. Most horses settle quite quickly

with this method and then are more willing to stand still for the awkward bits to be done. Care must be taken not to allow the horse to become tangled up in the wires. Clippers that run from a battery strapped to the waist are a must for this type of horse. But if a battery is not available and circumstances allow, the leads can be run along the roof of the stable and come down from the centre so that the clippers can be connected. This way, no leads are trailing on the floor for the horse to trample. Ideally the cables should be placed through loops so they are off the floor. Most modern clippers have connections that come apart quite easily so that if the horse does become tangled, the leads will separate and this will also switch off the motor.

There are other ways of restraining horses but these are best left to the professionals as they can be very dangerous if things go wrong. Someone who is very experienced will have the knowledge to see potential problems and prevent accidents.

USING COMPLEMENTARY THERAPIES

My favourite way of helping horses and ponies cope with the potential distress and trauma of clipping is to use some form of complementary therapy. By becoming familiar with these and using them effectively, the need for any kind of restraint is rarely necessary. Luckily, there are more and more veterinary surgeons becoming aware of the benefits of complementary therapies and they can refer you to a reputable practitioner. If not, the governing associations of these therapies can be contacted for further information and lists of qualified practitioners. A few vets actually use alternative therapies in their practice and are able to give owners valuable advice on their use.

Although this section deals with just one problem that can occur when dealing with horses, they can be extremely beneficial in all aspects of equine care. Complementary therapies, however, do not replace good horse management or skilled veterinary care. They do, however, work very well in conjunction with conventional veterinary treatment. Under the 1966 Veterinary Surgeons Act, it is illegal for anyone other than a veterinary surgeon to treat an animal, so your vet should always be consulted before embarking on any complementary therapy. This restriction has been made because animals have little choice in their treatment so their welfare has, quite rightly, to be legally protected. Your vet, by law, still retains overall clinical responsibility for a horse's treatment.

Bach Flower Remedies

One such natural therapy, rapidly gaining popularity amongst horse owners, is the collection of Bach Flower Remedies. There are thirty-eight remedies in all and each relates to a particular personality character trait and state of mind. The horse or pony may have psychological problems and his attitude and reaction to certain situations or life in general may be showing the negative side of his personality. Apart from finding the horse's 'constitutional remedy' which will help to balance the horse's emotions,

there are remedies that can be used as 'helpers' in times of transient traumatic emotional states which manifest while a particular activity is in process such as clipping.

Some of the remedies that may be beneficial in a clipping situation are listed below, but this is only a general guide and knowing your horse or pony well will help you choose the most suitable remedy or combination of remedies.

Mimulus – can be used for the horse in whom the cause of his fear is recognised, e.g. the noise and feel of the clippers. The horse has not had a very good experience with the clippers on a previous occasion and remembers this well.

Rock Rose – is good for horses who show extreme terror, anxiety and panic. These are the ones who sweat up quickly, throw themselves around the stable and are in too much of a state to listen to the handler's attempts at reassurance.

Impatiens – for horses who become 'stroppy', irritable and sometimes retaliate if they do not want to be clipped.

Vine – can benefit horses who are very strong willed, unruly, dominant and always want to be the boss, including when you clip them.

Star of Bethlehem – helps the horse to overcome the shock and trauma that can result from clipping.

Walnut – assists a horse to adapt to a change such as losing its coat, or a change of routine if brought in from the field to being stabled managed.

Willow – horses and ponies who resent being clipped and seem to 'sulk' and feel sorry for themselves afterwards.

Bach Rescue Remedy is a combination of five remedies: Rock Rose, Cherry Plum, Clematis, Star of Bethlehem and Impatiens.

Rescue Remedy does exactly what its name suggests – it 'rescues' the horse or pony from a traumatic experience. During a crisis or trauma, it acts to calm the emotions and bring about equilibrium again. It is excellent when given to horses and ponies who become upset while doing a particular activity and can be beneficial if given during clipping.

Rescue Remedy comes in liquid concentration or cream. The cream is very good when used on areas where the blades have accidentally nicked the horse or there are raised areas caused by the blades pulling at the coat or an allergic reaction to the blade washes and oils. Rescue Remedy reduces the trauma to a localised area and promotes self healing. The dose of Rescue Remedy for equines is ten drops in a bucket of water or four drops on a lump of sugar if the horse is not interested in drinking water.

Homoeopathy

Homoeopathy is based on the principle that a substance which produces a set of symptoms in a healthy body will cure a similar set of symptoms in a diseased body. It is a case of 'like curing like'. A homoeopathic remedy is not just chosen according to the diagnosis of a disease but to an individual's demeanour and disposition as well as the symptoms. For example: whilst two horses may contract laminitis, the symptoms are not necessarily identical. With clipping, the remedy is chosen depending on what emotional and physical response the horse shows. A horse who may be very afraid, panics and sweats up easily would require a different remedy from that prescribed for the horse who is afraid but becomes very aggressive.

Aromatherapy

Essential oils have been used since ancient times in Egypt, Greece, Persia, China, Rome, India and more recently in Europe. Like so many other complementary therapies, aromatherapy can be beneficial to animals, although at present there is not that much documented evidence. There are a few 'pioneers' working in this field and with good results; hopefully more will be known in the future.

The best known and most commonly used oil is lavender, which in humans is excellent for helping to induce relaxation and relieve nervous tension. When used with horses it has been observed to have the same effect. Some of the uses for lavender are: eliminating fear, panic, anxiety, worry, mood swings, restlessness, impatience and irritability – all the emotions that horses can exhibit when being clipped. A few drops of lavender can be rubbed into the base or tip of the ears or put onto a grooming mitt and ran all over the horse's body as if grooming and massaging, which helps balance the body's subtle energy flow. This can be done before clipping to relax the horse and may be used after it if unfortunately he has become upset, to rebalance the emotions. Lavender can also be put onto a cloth and inhaled so long as the horse is not upset by this; but most usually do not mind.

Radionics

Radionics (or the 'black box' as it was more commonly known) is a recognised complementary therapy. It can be described as 'holistic distant healing'. Holistic meaning 'whole', and 'distant' meaning that a radionic practitioner does not have to see the animal for any therapeutic action to occur. All that is required is a small piece of mane to act as a link between horse and practitioner and details of the problem.

Radionics is based on the belief that living beings are a complex, multi-dimensional mass of interrelating energy fields and everything that exists is energy in varying degrees of density. Each form of energy has its own vibrational frequency and if any one of these energy fields is disrupted then disharmony and disease – be it physical or psychological – will occur.

The aim of radionic therapy is to balance these disrupted energy fields by directing corrective energy patterns to promote the healing process, thus restoring equilibrium. With clipping, radionics seeks to eliminate negative emotional responses and reduce the stress and trauma by keeping the energy systems in balance.

Communication

Very often we create problems with our horses by failing to communicate with them – as a result misunderstandings and bad behaviour often arise. Effective communication is vitally important in our daily contact with horses. We rarely take time to communicate with our horses and take it for granted that their wishes are the same as ours. This, of course, is not always the case and problems inevitably occur as a result.

There are few horses who accept clipping without any resistance, no matter how small and seemingly insignificant. Clipping really is an intrusion and a very invasive action. Like everything we humans do with horses, we are taking a very great liberty when we clip them. It is only by their good nature and willingness to serve mankind that we are able to do anything with them at all.

Most communication is not done at a conscious level. A horse's way of communication is not one of speech, but one of thought and feeling. Verbal language is not how the horse communicates, although by being with humans, they learn to associate our spoken word and body language with what they are picking up from us. Horses receive our thoughts and intentions before a verbal request or riding aid is given.

We all know that gaining a horse's trust is important, and even more so if we are intending to do something he dislikes or is fearful about. If a horse is nervous or frightened then he is more likely to become calm and relaxed if he has trust and faith in his handler or rider. This is especially the case with clipping – so establishing a rapport is the first thing you must do with any horse.

When you clip, take time to explain to the horse about clipping, why it is necessary and what is going to happen. This can be done mentally or verbally, whichever you feel more comfortable with. When clipping a strange horse, make sure that you introduce yourself – there is much to be gained by being polite! By explaining to the horse what is going to happen, you may be amazed at the result. Because horses tune into us easily, be aware of sending clear, positive reassuring 'signals'. Our signals can often be scrambled, negative and confusing so it is not surprising that he doesn't understand what is going to happen. Be very precise, focus on the horse and what you are doing.

There is great emphasis on visualisation in riding training methods and the use of visualisation need not only be employed while riding. During clipping, visualise everything going well, such as the horse behaving and standing still, the lines and the difficult areas being clipped easily. If you visualise everything going wrong, then more often than not it will.

•6•

DIY OR EMPLOY AN EXPERT?

LEARNING TO CLIP

Clipping horses can be very dangerous and potentially fatal – it is not an activity for the fainthearted. The first step is to watch horses being clipped and then when you feel confident ask if you can assist with a few quiet horses during clipping. When you actually want to try to clip, it is wise to ask an experienced person to show you how to start and give you 'hands on' practice with the clippers on a placid horse. If you are assisting, the person clipping may let you clip some of the easier places. There are many helpful hints and tricks of the trade that cannot be learnt from a book. Nothing can take the place of getting in there and having a go yourself. Don't worry about making mistakes and clipping too much off; although it seems like a disaster at the time, the coat will grow back (eventually)!

A description of how to clip the horse is outlined in the next chapter, and you can refer to it for guidance.

Make sure that the horse you use for practice is familiar with clipping. The horse must be extremely patient and not nasty in any way. His temperament must be very amiable, as he will have to stand still for long periods and not object to being pulled about while you work on awkward areas and try to get the lines right. Make sure he hasn't any ticklish places, also. An easy-going horse to practice on will increase your confidence and skill no end. Later you will be able to try more difficult clips and eventually horses who are more challenging behaviour-wise.

For the first time use a horse that has already had one clip and is due for a re-clip. This way, the lines are already there for you to follow so then you won't have to worry too much about getting the lines right, as well as trying to get used to handling the clippers. This clip should be much easier as there is not so much coat to remove and the horse is much cleaner. You will already know the horse's reactions to being clipped, so this should help with your confidence.

Although you will no doubt take great care around the sensitive areas, 'nicking' the horse does sometimes happen, even to the most competent, experienced person. Don't worry – it's not the end of the world. Stop the bleeding and clip another area because the horse may be reluctant to let you near the cut area for a while. When you have regained his confidence, and provided the injured place is not too sore, try again towards the end of the clip, or the next day when the cut has healed up and the horse has had time to forget (a few horses don't, however).

Don't try and clip your own horse for the first time if he is not completely happy with clipping, even if you feel you know him very well. Any upset will set him back in his attitude to clipping, and this won't do his or your confidence any good. Wait until you are more experienced, then give him a try.

Make sure even then that you have someone around who knows what he/she is doing so that he/she can bale you out if you get into difficulty or if things start getting out of hand. With enough practice, good preparation and an understanding of your horse's idiosyncrasies the chance of problems occurring will be reduced. Observe how the 'expert' solves any behavioural problems and employ these methods if the same occurs again (providing they are within your capabilities). They will usually work again.

....
Never attempt to clip any horse without another person present – this is dangerous and extremely foolhardy.
....

If you are going to do your own horse's reclip, you would probably have been involved in helping with the first clip, which should have been done by someone very experienced, so you will have a basic understanding of what's involved. Take this opportunity to ask questions about things you are not sure of. He or she might give you a go with the clippers if you explain that you are thinking of learning.

If you cannot find anyone to teach you, local Riding Clubs, equestrian centres and equine departments of agricultural colleges sometimes run courses or organise clipping demonstrations, so it may be worth finding out if any are being planned in your locality. Most instructors giving the lecture/demonstration will usually let you clip part of the horse to give you an idea of how the clippers feel when being used. This is also a very good opportunity to 'pick their brains' about overcoming any problems that your horse has with clipping.

FINDING SOMEONE TO CLIP YOUR HORSE

There are always adverts in the local horse press for people who are willing to clip horses for a charge, but it is very sensible to have someone personally recommended to you by somebody whose opinion you respect. If other horse owners in your area have already had their horses clipped, ask how the person hired handled their horse. Were they patient but firm, confident and experienced enough to overcome any difficulties, or was the end result achieved by brute force? The way the horse is handled is extremely important, especially if he is a 'first timer' or of a nervous or aggressive disposition. Many horses' attitude towards clipping has been ruined by being badly handled and having a terrifying experience, so choose your 'clipper' very carefully indeed. Observe the quality of the clip – are the lines straight, is there a smooth finish or are there any scraggy patches left on? This way you know what you are getting and can be absolutely sure your clipper will do a good job.

Saddlers, feed merchants and local riding establishments often have notice boards where people advertise for business. Get in touch with committee members of your local Pony or Riding Club because they usually have numerous contacts and know the equine services available in the locality. Blacksmiths are another good source of local information because by the very nature of their job they do 'get about a bit' and always know who's who. Riding instructors may do some freelance clipping, and professional grooms will also take on clipping work.

Costs can vary; there is always someone around who is prepared to clip your horse cheaply, but it can sometimes work out very expensive in the end. The quality of the clip may be poor and, more importantly, if they handle your horse badly they could ruin him for life. You may have to pay someone else to tidy up the clip, and regaining your horse's trust and confidence may take years and much patience. Sometimes badly treated horses never trust anyone completely again.

Professional people who charge exorbitant prices for clipping can also be guilty of upsetting horses to the point of making them unclippable, but the likelihood of this is small because they usually wish to preserve their good reputation and a 'bad

report' can affect their earnings. It is still sensible, however, to have even the professionals recommended to you.

Professional people who clip have overheads. They may seem to be charging a lot, but they have to pay for blade washes and oils, spare fuses, masks, overalls, regrinding blades (which can blunt very quickly on dirty, greasy horses), maintenance of clippers, servicing, replacing cables if trodden on, replacing clippers if they are damaged, and buying new blades if broken. Some people are still grossly overpriced, so it always pays to 'shop around'.

Clipping takes time, patience and experience so, when pricing clips, the skill involved must be taken into consideration. The person doing the clipping is putting themselves at great risk, even with quiet, sensible horses, so it is only fair that they make it worth their while. Sometimes they add 'danger money' to the price of the clip if the horse is extremely difficult. Some people charge a standard rate for clipping whichever clip you decide upon. Others have a range of prices depending on the amount of coat to be removed or the difficulty of the clip. If they have to travel some distance then expenses for petrol may be added on to the price of the clip, so enquire about this beforehand. Always obtain a full quote so you don't have any added surprises on the day.

When contacting the person you feel is best suited to clip your horse, make sure that you are honest with them. If the horse hasn't been clipped before, or if he is nervous, slightly difficult or downright dangerous – please tell them. There is absolutely nothing to be gained by saying

Watching a professional in action is the first step in learning to clip.

he is an angel when in reality he is a lunatic! This way they know what to expect and will arrive well prepared for problems. This will reduce the likelihood of anyone getting injured (including the horse). Remember that if the person clipping is injured, and the injury is severe enough, they may be unable to work for some time. For grooms and people in low-paid jobs, any time off may cause real hardship and could even lead to dismissal. Check if their and your insurance policy covers injury to people or damage to clipping equipment, just in case of an accident. Even with the quietest pony, best preparation, experienced clippers and handlers, accidents can still happen.

Also be honest about your own ability to hold your horse and assist during clipping. If you are very nervous and apprehensive about having your horse clipped, then he may become restless beforehand and upset during the actual clipping. It is well known that horses are very good at tuning into peoples' emotions, especially their owners. Ask someone else on the yard, who you know is confident, to hold your horse instead. If no one is available then let the person clipping know that you are worried. They will usually be able to bring an experienced assistant with them. Many horses I know behave better without the owner around!

If you are unsure of what is required for clipping, ask your clipper to give you advice on things to do beforehand, such as getting the horse clean, preparing the stable and other items that are needed such as rugs, haynets, bandages etc. Ask if there is anything else you should provide, such as extension cables, something to stand on, tea and biscuits! If you are not sure which clip to choose, he or she can also advise you on this if you provide details of the type of horse, work expected and type of management through the winter.

Make sure that everything is prepared in plenty of time before the person you have booked is due to arrive on the yard. They will have allowed time enough to complete the chosen clip and for any behavioural 'hold ups' on the horse's part, but not the time needed for getting the horse ready and preparing the environment. If this happens, they may be reluctant to clip your horse again if they are made late for other clients. If you decide to change the clip to one that takes more time, let the person know as soon as possible; don't surprise them on the day because they may have to 'shuffle' other people around in order to make more time for your horse. Give the person very detailed directions to the stables and also the yard number just in case they get lost (as I did frequently)! If the horse needs to be sedated by your vet, make sure all parties are aware of the time. It is essential that everyone is on the yard at the same time. It is no good having the horse sedated and ready if the person clipping is due to arrive half an hour later or has been held up.

Not everyone who provides clipping services is willing to travel, so you may have to hack or travel your horse to them. If hacking, allow plenty of time to cool, dry and brush the horse off. Arrange for someone to take rugs to the yard to put over him while he is cooling down and standing around waiting. Someone on the yard may lend you a rug, but don't count on it because they may be wary of spreading infection to their own horses. Take a headcollar or halter with you as these are much easier to clip around than a bridle. The same applies if you are travelling by box. Still allow plenty of time for your horse to cool and settle, especially if he tends to sweat up while travelling. If boxing, he may be 'fresher' than if you have hacked him there, so walk him around to get used to the unfamiliar environment so that he is more relaxed for clipping.

·7·

CLIPPING YOUR HORSE

EQUIPMENT REQUIRED FOR CLIPPING

○ Clippers.

○ Clipper blades – fine, medium, coarse, whichever is required for the clip.

○ Clipper oil – for cooling and lubricating clippers.

○ Blade wash – for cleaning and sterilising clipper blades (if indicated by manufacturer).

○ Small brush – to remove clogging hair from clipper blades and filter (a small paint brush is ideal).

○ Clean body brush or another soft brush.

○ A set of small, quiet, cordless clippers or hand clippers.

○ Long electric extension cable.

○ Power-breaker socket plug – to prevent shocks.

○ Spare fuses – should one blow.

○ Screwdriver – to change fuse.

○ Overalls to keep clothes clean (loose hair is less likely to stick to cotton overalls; nylon overalls can be a problem with static electricity).

○ Face mask – to safeguard against chest problems from horse hair and dust.

○ Headscarf or hat to keep hair out of the way and clean.

○ Rubber-soled footwear – preferably with steel toecaps.

○ Twitch – only to be used if horse is very difficult.

○ Scissors.

○ Spare tension screw and nut.

○ Cloth for wiping excess blade wash and oil off blades.

○ Container for blade wash (old ice-cream tub is ideal).

○ Rugs to put over horse.

○ Haynet – full.

○ Tail bandage.

○ Elastic bands.

○ Chalk or marker for guidelines – white for bays, chestnuts and blacks; and coloured marker for greys, and white areas of skewbalds and piebalds.

○ Sponge.

PREPARATION FOR CLIPPING

○ Find a sheltered clipping area, ideally

with plenty of natural light. Clipping can take between one and three hours. It is sensible to start in the morning so that you don't run out of daylight. If this is not possible ensure that the electric lighting is adequate.

❍ Prepare the clipping area by removing such items as water buckets, sharp objects that the horse may run into, and stable fittings that are potential conductors of electricity. If this proves to be totally impractical, any metal or sharp objects can at least be covered, preferably with rubber, to prevent shocks. If clipping in a garage or farm building, then ensure that the horse cannot damage himself on anything that is stored there if he misbehaves.

❍ Leave some bedding down or use rubber matting, so that the horse's foothold is secure.

❍ Make sure that the circuit-breaker is plugged in and has been previously checked to see that it is working.

❍ Check that the extension lead will easily reach from the socket to the clipping area without straining and stretching the cable.

❍ Ensure all that all the clipping equipment has been checked and is in good working order. The clippers themselves should be greased and well oiled, and the blades should be clean and sharp.

❍ Find something stable and secure to stand on if you need it, to reach the back or head of large horses.

❍ Organise a very experienced and capable assistant if one is necessary. In any event, make sure someone else is around.

❍ If possible, choose a time when the yard is fairly quiet and there is not too much activity.

❍ Know where the first-aid kit is kept – both for horse and human.

❍ Make sure that the horse or pony is very clean. He may need to be bathed the day before clipping, but see that he is kept warm and dried off quickly and thoroughly afterwards. Some horses' coats are extremely greasy and this is caused by the natural oil called sebum, which is secreted by the glands in the skin. The amount secreted is increased during the winter and helps to make the coat waterproof, but it can clog the blades if it is excessive.

If you don't have the facilities for bathing the horse, then the worst of the grease can be removed by sponging the horse's coat. Put some warm water into a bucket, then with a wet sponge, sponge a small area working against the lay of the coat. Clean the sponge in the water and go over the same place if more grease needs to be removed. If this area is fairly free from grease, dry with a towel and move onto another area. Be careful not to have the sponge saturated with water. It may be a good idea to add a mild shampoo to the water to help break down the grease so that it can be removed more easily.

❍ Clip on a mild day if possible, not a very cold windy day. Clipping on a freezing cold day may come as a bit of a shock to the horse's system! Remember that you are removing the horse's natural insulation and he will need time to adjust to this, even with favourable conditions and good management.

❍ Have everything close to hand such as rugs, protective boots and bandages, full haynet, tail bandage, brushes, elastic bands, titbits to reward the horse, etc. Make sure the rugs have been aired and any tears, broken straps and buckles have been mended or replaced.

❍ The mane may need to be plaited to keep it out of the way of the blades and to avoid chunks being clipped off by mistake. Make sure that the plaits are not too tight as this will pull the skin and distort the line along the base of the mane if all the neck is being clipped off.

When clipping near the mane be careful not to clip too close as some of the mane will be removed. This then grows back very 'spiky', looks awful and can take a long time to re-grow to the original length of the mane. Some horses have hair growing along the mane line that is not coat, but not exactly mane either. If your horse has this type of hair growth, it can act as a natural marker, so don't go any nearer than this.

❍ A bandage should be put on top of the tail so that you can easily clip the area by the dock on a trace, blanket, hunter or full clip without cutting into the dock or clipping the tail hairs off accidentally. The rest of the tail can be plaited below the tail bandage so clipping around the back of the horse will be easier. If left loose, it would be much more difficult to

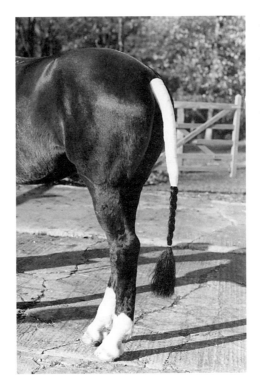

Bandaging the tail prevents the loose tail hairs being clipped off accidentally. Plaiting the tail all the way down makes it easier to move the tail out of the way when clipping the hind legs.

manoeuvre the tail out of the way to clip around the horse's back legs.

❍ Long travelling boots or stable bandages can be put on if you wish. Knee and hock boots may also be necessary. Protecting the legs is best done as a precaution if you are not sure how the horse will react or if he has been sedated.

❍ If you are not yet experienced enough to be absolutely sure where to make the lines on a trace, blanket or hunter clip by eye alone, then mark off these areas with chalk or tape. The illustrations and descriptions of the various clips in Chapter 2 will help you to decide where the lines should be marked off.

❍ Clipping a horse is hard work, so

wear comfortable loose-fitting clothing. Don't wear a lot of layers as you can quickly become hot, especially if you are dodging out of the way for most of the time! Wool is not very good to wear because the hair will stick to it; nylon causes static; cotton overalls are ideal. Wearing a mask saves a lot of damage being done to the lungs from dust and hair. This is essential if you intend to clip many horses through the winter.

❍ Adjust the tension of the blades as indicated in the manufacturer's instructions, and oil the blades before starting to prevent 'snagging' of the coat if it is clean and the blades are sharp.

❍ Remove your watch before putting the safety strap over your wrist; this will prevent the strap snagging on the watch.

THE CLIPPING PROCESS

A step-by-step guide to the actual process of clipping is difficult to describe because you have to take into account the style of clip and the individual horse or pony. For the most part, you have to be guided by the horse's responses and adapt your clipping sequence to him. It is no good starting on the most common place, which is usually the shoulder, if he does not like this. Don't have a battle to start on the shoulder if he is quite willing for you to start on his side or his quarters. Observe the horse's reaction as you approach a certain area of his body, and take your cue from him. From the illustrations/photographs and description of the clips, you should have some idea of what part of the coat is to be removed.

A few points have already been mentioned previously, but they are very important if you want a good quality clip and an unspoiled horse for clipping.

❍ Describing where to stand is difficult as it depends on what part of the horse you are clipping. Never put yourself at risk. Common sense should really dictate your position in relation to the horse's danger areas. It is asking for trouble if you stand directly behind the horse where he can easily kick you, or immediately in front where he is likely to strike out with his front foot. Always stand to the side of the horse and use the length of your arm to reach the difficult places. If you bend down, the horse may catch you in the face if you are close to the legs. Being cautious about where you stand does not mean that the horse cannot catch you, but there is no point in making things easy for him.

Even if you have managed to put most of the cable through loops so it is not dragging along the floor, you still need a certain amount of 'slack' so that you can reach the horse and manoeuvre the clippers easily. Make sure that the slack is not so long that the leads touch the floor; if they do, then you are in danger of tripping over the cable. So that you don't have cables trailing around the horse's legs, turn the horse around to reach the places that need clipping,

❍ Let the horse nibble at his haynet; this will keep him occupied while being clipped. Keep titbits nearby to reward him when he behaves well.

❍ Switch the clippers on away from the horse and watch his response. If you already know the horse very well you may be able to go straight up to him and start. But, if the horse is nervous or young then take time to familiarise him to the noise and feel of the clippers. This may have to be done over a period of days or weeks but it is a stage that should never be rushed. How best to do this is outlined in the section on clipping difficult horses. Talk to the horse and reassure him all the time, even if he has been clipped many times before.

❍ Start clipping wherever he will let you, probably at the shoulder or base of the neck. Approach the horse in a calm, confident, positive manner. 'Old hands' and 'stroppy' horses will recognise any hesitation and nervousness, both in body movements and tone of voice. They can sometimes use this to their advantage! Don't try to compensate for being nervous and apprehensive by being over domineering and loud. Horses 'tune into' peoples' minute body movements

Placing the blades on the back of your hand with the palm lying flat against the horse's body, can help the novice horse to become accustomed to the strange sensation of the clippers. Watch the horse for any signs of distress or annoyance. Reassure him to let him know that there is nothing to worry about.

This tufting is a fault caused by uneven pressure of the clippers on the horse's body.

Make sure that each sweep of the blades overlaps slightly or you may end up with something like this.

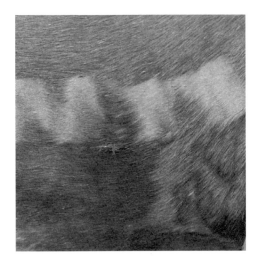

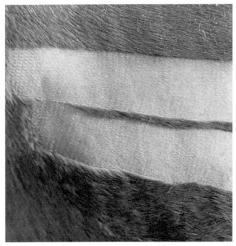

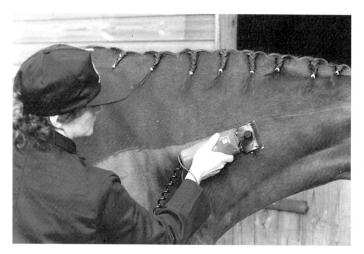

Clipping up the neck. Once the line has been marked out, it is very easy to follow.

BELOW: *Pulling the leg forwards allows you to get closer to the elbow and makes the area flatter and firmer.*

very well as they do with other horses, and are not fooled by our over-compensatory behaviour. As previously mentioned, it is well known that horses can 'pick up' human signals of emotional states, even if we are not consciously aware that we are sending any out. Don't put up with any nonsense, however, even if you are slightly nervous and are aware that the horse knows this.

○ The clippers must be held against the horse with the blades nearly flat. They should be kept on the horse with a constant, firm pressure but not digging in.

Remember that the skin is very sensitive. Don't force the blades through the coat, let them glide along; the weight of the clippers should take them through.

○ Clip against the lie of the coat and observe how the coat can change direction. The coat grows in certain ways to channel any wet downwards, off the horse's body. Notice that the bulk of the horse's coat is very dense, which helps to insulate the horse against the cold. This dense coat is interspersed with longer, finer hairs that help to drain the rain off the horse's body to prevent him becoming saturated through to the skin. The coat can have other peculiarities such as whorls and wheatsheafs, and the hair under the stomach changes direction quite frequently. You have to be very adaptable with the angle of the clipper head, so that the direction of the coat is followed and no tufts are left.

○ Clip off a strip of coat, and make the next stroke parallel to the one before. This can be above or below, depending on the part of the horse you are clipping but it must slightly overlap the previous

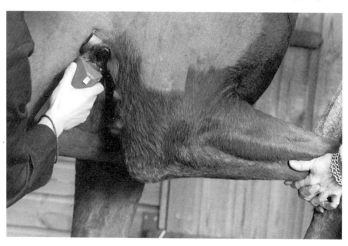

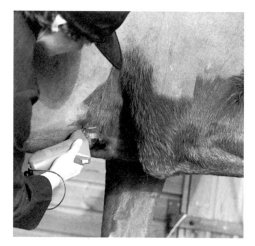

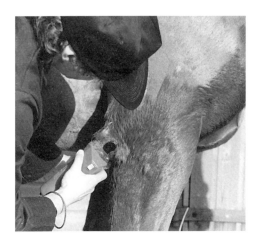

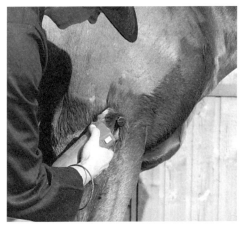

LEFT: Follow the line of the muscle to achieve a good angle at the top of the foreleg.

Tidy up this area before finishing off behind the elbow.

stroke. Long, sweeping strokes will usually prevent marks or patches of hair being left on.

If the coat does happen to be dirty or very thick, then shorter strokes are necessary to prevent clogging the blades. The same area may have to be clipped a few times, each cut reducing the thickness of the coat until it is the required length. An extremely thick coat can overload the blades with hair; this happens even with the most powerful clippers. If this occurs because the coat is very thick or greasy, the horse can first be clipped with a coarse set of blades, then brushed or washed thoroughly and the clip finished off with a medium or fine set of blades. Remember to keep a rug over the horse, especially the areas that have already been clipped. The rug can be moved around to expose the areas you are working on, while keeping the clipped areas warm.

○ Be very careful not to cut the horse when attempting the awkward places such as between the front legs. The skin here is very thin on most horses and easily cut by the blades. Ask an assistant to

hold the front leg up and forwards so that the folds of skin are stretched to make the surface flatter and firmer. Hold the leg above the knee joint so as not to pull the tendons. With very narrow horses and ponies, the leg can be moved out to the side very slightly, so that the blades can get between the legs.

Make sure that the horse is comfortable about holding his leg in this position. Some older, stiff horses may not like their leg held up for too long. Try doing this area in stages, as having a battle will only upset the horse. With your free hand, firmly but gently pull the skin that isn't flattened by pulling the leg for-

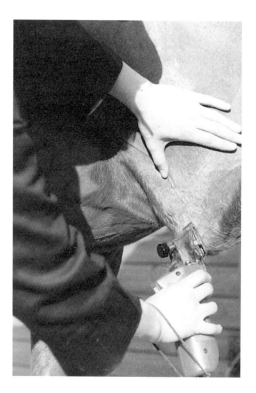

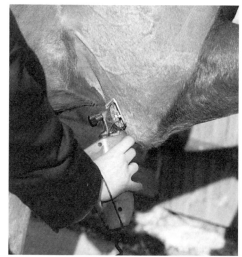

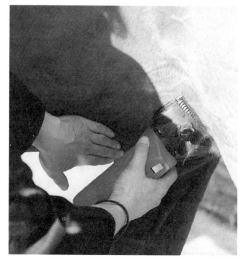

Again, pulling the leg forward will open up this area making it easier to clip. Flattening the skin with your free hand (see above and right) stretches the skin a little more so that loose skin doesn't ride up in front of the blades.

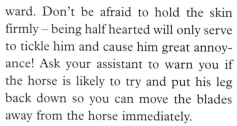

ward. Don't be afraid to hold the skin firmly – being half hearted will only serve to tickle him and cause him great annoyance! Ask your assistant to warn you if the horse is likely to try and put his leg back down so you can move the blades away from the horse immediately.

If the natural light is fading, and electric light is not bright enough, don't attempt to clip around the elbows and between the legs as this increases the risk of nicking the horse. Finish the clip

another day or get another helper to hold a very bright torch and shine the beam onto this very difficult place.

❍ On the outside of the foreleg at the top, there is a line of muscle that can be followed for getting the line right. The line on the inside of the leg should correspond with this. Unfortunately, the line on the hind leg is not so obvious but a general rule is that the start of the line is in the middle of the curve on the back leg from the buttock to the point of the hock. This line should be taken upward and finished a few inches below the stifle at the front of the leg. Again a corresponding line should be clipped on the inside of the leg. This can sometimes be reached more easily from the other side of the horse, which lessens the likelihood of being kicked. Put your hand just above the hock and hold it firmly; this may discourage the horse from kicking you. By doing this, you will be able to feel if he

In the areas where the skin is loose, it can ride up in front of the blades and is quite easily cut. Pulling the skin with your free hand makes it flatter and firmer and therefore easier to clip. The risk of nicking the horse is greatly reduced.

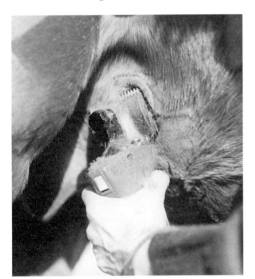

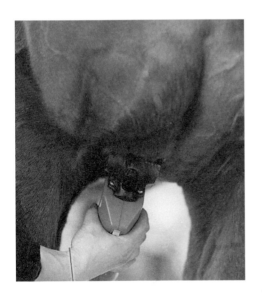

Apart from the head, between the front legs is one of the most difficult places on the horse to clip and the site where the horse is most frequently cut.

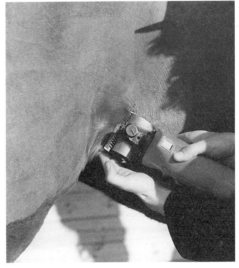

Putting your hand behind the loose skin in front of the stifle makes the area easier to clip, but watch out for horses who may be ticklish in this area. Hold the skin firmly.

These two photographs show the angle of the line on the back legs. Be very careful where you stand as the horse can easily kick you when you are clipping this area.

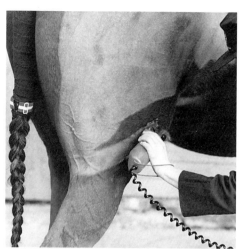

makes any attempt to do so and move out of the way quickly. Make sure that the two 'vees' on the back legs are level with each other. The points on the front of the forelegs must also be equal.

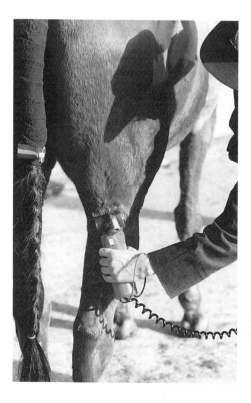

Don't go any lower than this when starting to clip the 'rear end'.

○ When clipping the belly, watch out for the coat changing direction, and 'cow kicking' horses of course! Some horses are very ticklish in this area. Be careful not to cut the sheath of a gelding or stallion or the udders on a mare – they certainly won't let you near them again if you do. This can happen if the horse is very hairy in this region.

Another place that is at risk of being cut is the soft skin in front of the stifle which is also attached to the bottom of the flank. The coat also grows in different directions here, which makes it all the more difficult to clip. If the horse is not too ticklish, use your free hand behind the skin to steady it and stop it from 'wobbling' about.

○ If clipping the legs, then do the top half from the knee or the hock upwards first. When clipping the lower leg, be very careful around the fetlock as there are small horny growths there called 'ergots' and the blades can quite easily cut into these. The area around the chestnuts should also be clipped with caution, especially if they have been allowed to

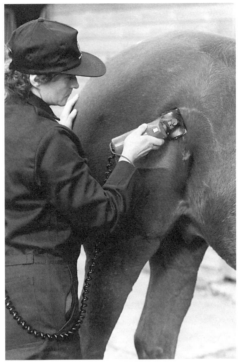

Clipping out the 'arc' on a trace or blanket clip. With practice the arc can be clipped out in one smooth movement, as shown in this sequence of photos.

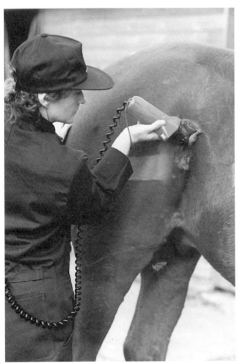

When clipping along the base of the mane, be careful not to clip off any of the mane by mistake.

become overgrown. (The chestnuts are found on the inside of the back legs below the hock and on the inside of the front legs, above the knee.)

The heels can be made more accessible by holding up the leg at the fetlock.

Putting a rug over the horse's back and quarters prevents him from becoming cold when these areas have been clipped. Fold back the rug to reach areas not yet clipped or that you want to go over again and replace the rug when finished.

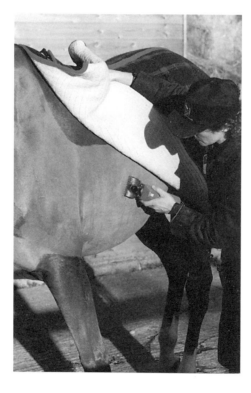

This opens up the area but only works if the horse is relaxed or not trying to pull away from you. Holding up the leg also serves to 'flatten' the skin on the lower part of the leg so that it is easier to clip and not ridged.

❍ If attempting a hunter clip and clipping around the outline of the horse's own saddle, leave a margin of about 5cm (approx. 2 inches) so that you have some coat to play with when tidying up the shape and neatening the edges. Check that the patch is equal on both sides of the spine and that the two halves meet at the same point at the withers. Also, leave a triangle at the top of the tail (see photograph on page 22). The point is towards the horse's head, along the path of the spine.

❍ If a blanket or trace clip is being done, remember to clip an 'arc' out of the line over the stifle area (see previous page).

❍ The head is probably the place where you need the most patience and must take your time. Only go as fast as the horse will let you. Some form of mild restraint or sedation may be helpful if the horse does not respond to reassurance and firm, sympathetic handling. A great deal of damage can be done to a horse who throws his head around. An eye can be badly bruised or cut if knocked by the blades. Leave a small area of coat around the eye if the horse or pony is not happy about you clipping this area. This is better than trying to clip so close as to put the eye at risk.

Sometimes, if the horse is very frightened and won't accept the clippers on

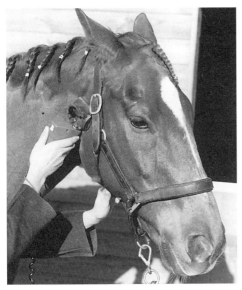

Clip as close to the headcollar as you can
before you have to remove it to clip the
remainder of the head. Clipping in between
the headcollar straps can be dangerous as
the clippers can be caught up.

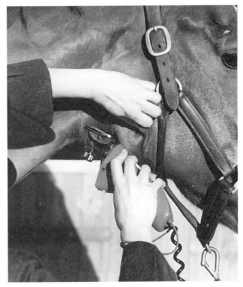

Pulling the skin tighter around the throat
makes this area easier to clip.

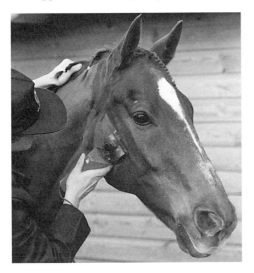

After removing the headcollar, begin on the
cheek as this is the place the horse is least
likely to object to. Carefully and slowly put
the clippers on the cheek – switched off
initially. If necessary spend time reassuring
the horse before you begin.

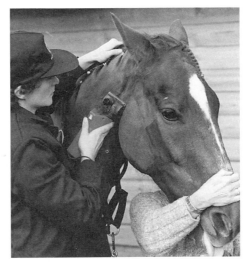

Tidying up below the ears.

the head under any circumstances, it is safer for horse and handler to leave the head unclipped. If this is the case, finish off tidily around the throat and as close to the ears as you can, possibly finishing this area with small cordless clippers or even handclippers. In any case, the head doesn't sweat an awful lot so leaving the head hair on makes little difference to keeping the horse cool and it does serve to save a great deal of warmth. A few horses will let you clip the hair from the bottom half of the head but not the top half, above the line of the cheekpieces. If the rest of the horse has been clipped before the head is started, make sure he has been rugged up properly so he doesn't become cold and start to fidget.

○ The head has many bony prominences and grooves. It takes great skill to clip the head tidily without hurting the horse. Loose skin under the chin up to the throat can be stretched slightly with your free hand so that it is easier to clip. Your assistant can raise the horse's head so you can see this area better. Watch out for loose hair falling into your eyes if you are clipping a large horse.

Gently but firmly hold the ears and clip along the outside edges. Do not clip the hair on the inside of the ear as this offers protection against flies, parasites and debris.

Shielding the eye with your hand prevents any loose hairs falling into the eye and can help horses who worry about the clippers being close to their eyes.

A horse's ears may need a lot of gentle manipulation if the hair around the ear and poll region is to be clipped off tidily.

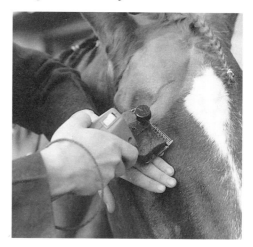

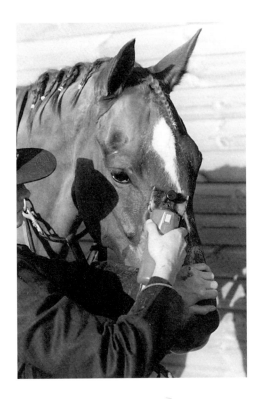

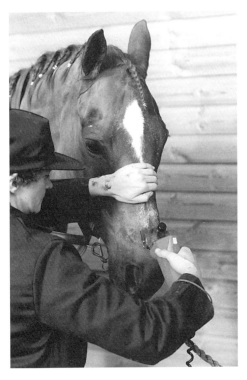

These three photos show how to clip the front of the face, using your free hand to keep the horse still. You may need the help of an assistant to hold the head of a horse who does not behave as well as this one.

To get at the head better, the head collar can be taken off and fastened around the neck, but the noseband must be facing backwards to stop it getting in the way. One that unbuckles at the nose is useful. Your assistant can hold the head steady while you clip. If removing the headcollar is not practical because the horse is not behaving well, a halter can be used to keep some control and is easier to work around than a headcollar.

The long hairs around the muzzle are 'sensors', used to feel and test objects. It is therefore much kinder to leave these on but some people prefer to remove them for cosmetic purposes.

❍ The hollows above the eyes are difficult to clip because some are quite deep, especially on horses who lack condition (and these should not really be

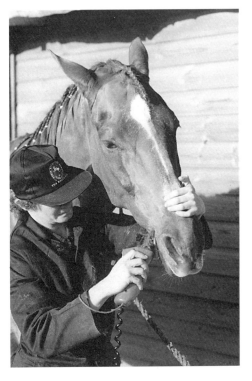

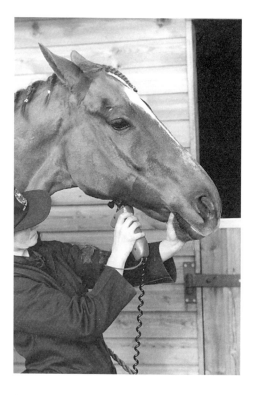

To clip under the chin, lift up the horse's head by putting your hand in the chin groove. This also helps to keep the head still.

clipped anyway) and older horses. This area must be clipped very gently as the vibration from the clippers can be felt intensely on the bone surrounding the hollows. The skin immediately outside the hollow can be pulled down which acts to 'lift' the skin up out of the hollow so the hair can be removed. This is much kinder than trying to 'dig about' in the dip.

A small cordless set of clippers may have to be used for the head if the horse is quite nervous or for a pony with a very small head. Cup your hand over the eyes so that loose hair does not fall into them and upset the horse. Never clip

off the eyelashes.

○ If the horse does happen to be dirty and after clipping there are 'tufts' visible, brush or sponge the horse over thoroughly and run the clippers over the horse again. Sometimes the clippers will leave 'tramlines' even though everything has seemingly been done to prevent this. They will usually disappear over the next few days. This is where the blades have managed to catch the horse's hair and pulled the skin.

○ Throughout the time it takes you to clip, **remember to cool, clean and oil the clippers frequently,** usually every ten to fifteen minutes. This may need to be done more often depending on the thickness of the coat and how greasy the horse is. This point cannot be stressed often enough. Follow the maker's instructions for the particular machine you are using.

○ Every five minutes check if the blades are becoming hot by feeling the underside of the clipper blades. It is a scientific fact that two surfaces of metal rubbing together will produce heat, even if the manufacturers claim that they are 'cool running'. It is better to prevent the clippers becoming hot in the first place than to let them overheat and burn the horse.

○ Always brush away the loose hair from the blades and clipper head before using the blade wash and oiling the clippers. This activity also gives you and the horse a break so that you don't become tired and the horse does not become restless. If the clippers become hot very quickly, make sure the filter is not clogged with hair and dirt or re-adjust the tension screw.

❍ Once or twice during clipping, take the blades off completely and clean the clipper head and blades thoroughly. This will remove any hair and grease that the blade wash and oils have been unable to clear. Make sure that all the pieces are kept together in order so that you can reassemble the clippers easily. The tension screw, spring and nut can be lost in the bedding so put them onto a clean surface such as a cloth or piece of paper so they can be easily seen.

❍ After you have finished clipping,

The head completely clipped out on a hunter or full clip.

remove the remaining dust, grease and clipped-off loose hair with a soft brush or damp warm sponge/cloth. A very small amount of mild shampoo can be added to the water to break up the grease so that it can be removed more readily. Loose hair can stick to the inside of the stable rug and make the horse uncomfortable. He may even try to get the rug off if the irritation becomes unbearable. If the horse does begin to tear at his rug after he has been clipped it may well be that he is allergic to the material the rug is lined with. Rugs with a high proportion of nylon can produce static which causes the same 'prickly' effect that we humans experience. An untreated cotton sheet can be used to line the rug to prevent this.

❍ Take out any plaits from the mane, and remove the tail bandage and leg bandages if used. Rug him up warmly and put him into his own stable, which should have been bedded down in readiness. If you have to use the horse's own stable to do the clipping, any mess such as clipped-off hair, manure and soiled bedding must be removed. The hair will not rot down if put onto the manure heap, so try and separate this from the bedding and dispose of it with the household or yard rubbish. For a really smart appearance with a permanently stabled horse, over the next few days, add the finishing touches by pulling the mane, trimming the tail and other 'hairy areas'.

❍ Clean the clippers and blades thoroughly before putting them away. Remove the blades, clean and wash them in a dilute antiseptic and/or disinfectant solution (such as Savlon) to prevent spreading infection. Send the blades away for regrinding if they have become blunted. Clean the hair off the filter if your set has one. Oil the clipper head so that it does not go rusty while in storage (most modern clipperheads are made of non-rust material and are coated to prevent this). Store the clippers in a safe, dry place where they will not be knocked and broken. Check the cables for any damage.

.8.

MANAGEMENT OF CLIPPED HORSES

Because the horse's natural protection has been removed, it is therefore necessary to keep him warm by artificial means. This means using stables and field shelters, rugs, concentrate feed and hay, bedding and bandages. There are many excellent books around on stable and field management that cater for the complete care of the horse. The following relates to some aspects of management that are applicable to all horses but could directly affect the comfort and well-being of a clipped horse. With all the various types of stabling, rugs, bedding, feeds and nutritional advice available today, there is no excuse for a clipped horse, living in or out, to lose condition and become ill.

Good management will also reduce the number of times that you have to clip. The coat of a warm, healthy, happy horse will not grow back in so quickly. Remember that the coat regrows from the moment it is removed, the only thing that we can do is to try to reduce the growth rate as much as possible. It is much better to keep the coat short throughout than to let it grow back quite

long and then reclip. The horse will not have to re-adjust to losing the coat again, as he had to for the first clip.

Take time to watch your horse everyday so you know what is normal for him and you can more easily detect any deviation or decline in condition. This will show in his coat and behaviour.

Know your horse's normal temperature, pulse and respiration rates. Sometimes after clipping, the horse may go down with an infection or illness. Whilst the clip will not directly cause the problem, if the horse is slightly under the weather beforehand, then the stress of clipping may tip the balance in favour of dis-ease.

Do not clip immediately after moving to new stables. The horse will be going through a period of re-adjustment and under quite enough stress for the moment. His delicate immune system will also be exposed to all the 'new' bugs that he has not previously come in to contact with and it is well known that new horses on a yard frequently go down with a mild infection. The trauma of clipping will only serve to overload an already hard-working immune system.

To adjust the horse's management to conserve warmth, it is important to recognise when a horse is cold. An obvious sign is the coat standing up and looking 'starey'. This is how the coat traps a layer of warm air for insulation. If a horse has to do this then you can be sure that he is cold. Shivering is another obvious sign, and horses that have become wet and cold can be seen shivering to help maintain body temperature. This depletes the body of valuable energy and ultimately leads to loss of condition.

Feeling the horse's ears is a good indicator of his temperature if you are otherwise not quite sure. The ears should be felt with your bare hands (i.e. not while wearing gloves). If you feel that the ears are cold at the top only, then he is only slightly cold. If the base of the ears are also cold then he is very chilly indeed and action is needed to prevent him becoming hypothermic.

Also check the belly, especially the area underneath the rug, in front of the sheath/udders. If these places are cold then reassess your management of the horse.

Think initially about appropriate ways of increasing the horse's temperature. This could mean putting on extra rugs or blankets, dry rugs, thatching with straw, rubbing down with towels or a wisp to stimulate circulation and providing a warm feed and hay. Afterwards make sure that correct future care is given to maintain body warmth at a constant level.

If the horse starts growing long 'cat hairs' this is another factor that points towards the horse being cold.

STABLING

The horse evolved on extensive grassy plains, and for survival in its natural environment is designed to be 'on the move' and to escape predators swiftly. Being confined in a small enclosure, whether it is a field or stable, is totally alien and about as far from a horse's evolutionary function as you can get. Freedom of movement is essential to a horse's well being, both physically and mentally. By being mobile, the horse is able to keep himself warm, just as we can keep active in freezing weather to stop becoming cold. It is essential for a horse or pony to have a stable large enough to allow free movement to help him keep warm as well as preventing him becoming stiff.

The size of the stable is very important. For a large horse, a box measuring 3.5m x 3.5m (12ft x 12ft) is the recommended size; 3.5m x 3m (12ft x 10ft) is just enough for a small horse (around 15.2 hh) and 3m x 3m (10ft x 10ft) is only suitable for a pony.

The position of the stable is very important. Make sure that the wind and rain cannot blow directly into it, as this will dramatically reduce the overall temperature. In Great Britain, a south-facing stable is ideal. As well as rain blowing in through the door, leaks in the stable roof will make the bedding and floor very wet, which again will reduce the temperature.

Make sure that the stable is well ventilated but without draughts. If the stable is draughty then the horse is in danger of becoming very cold and chilled.

The top door of the stable should always be kept open, except in extreme weather conditions; then another method of ventilation must be made available.

Stables should therefore always have a window, ideally with bars or toughened glass so there is little risk to the horse of the glass being broken. The best type is designed to fall inwards, so that the incoming air is directed upwards. This prevents a draught because the fresh, heavier, cooler air is warmed and dispersed before it descends onto the horse's back.

Line the walls if there are gaps that the wind can blow through. Draughts can also enter the stable underneath the door if there is a gap. A draught under the door will ultimately make the stable colder and could even chill the horse if it is directly in line with where the horse lies down. This is even more of a problem if the stomach and legs are bare because of clipping. Put a piece of rubber or a plastic sheet across the bottom of the door. This will not interfere with the movement of the door when opening and closing. If this is not possible, then bring the bedding right up to the door so any draught is prevented from entering the stable by this access.

Some consideration must be given to the drainage, because the stable floor must be kept as dry as possible. If the stable is allowed to become wet and soggy then valuable heat will be conducted away and the stable will quickly become cold. This will eventually effect the horse's body heat. This is because water is an extremely good conductor of heat, and any warmth in the stable environment will be lost. The floor can be made from a non-absorbent material that dries fairly quickly, and slightly sloped forward so that the urine drains away from where the horse is standing. A wet floor can be very slippery depending on the material used, so if your stable has a slippery floor, leave bedding down to reduce the likelihood of the horse falling over and injuring himself.

Light is also very important, especially as it partly controls, along with temperature and food, the rate at which the coat grows. Because we wish to reduce the number of times we have to clip through the winter, it is very important that the stable is light as this will repress coat growth. The light should preferably be natural daylight, which isn't very long anyway in the winter months.

FIELD SHELTER

All field-kept horses and ponies, and those who are just turned out during the day, need some kind of shelter which will help to take the 'bite' out of the wind and/or rain. This may be natural shelter, such as a line of trees, a stone wall, high thick hedges, or hollows, gulleys and banks; or 'man-made', like backs of buildings or a purpose-built field shelter. For a clipped horse living out, a field shelter is essential and there are many good designs that are ready made and easy to erect. The purpose of the field shelter is to provide a dry, fairly warm and draught-free environment.

If you are thinking of making your own, then certain aspects have to be considered. The size depends on the number of horses that will be using it. The whole of one side needs to be open, which will allow plenty of fresh air but won't let the wind blow straight through. The position of the shelter is extremely important. It must have its back to the prevailing wind, as for stables. For effective drainage, it needs to be situated on the driest and highest part of the field. In very adverse weather, the field shelter can be adapted to make a temporary stable if a permanent one is not available, but in these times, a 'back-up' stable should always be kept in mind, just in case of extreme weather conditions, accidents and illness.

Avoid using construction materials that are good conductors of heat, such as metal or asbestos; wood is a much better material for insulation. A long hay rack that covers the length of the back wall should be fitted. This should be filled along its whole length so that if there are a few horses using the shelter, they will not fight over small areas of hay. By providing hay in the rack horses and ponies will be encouraged to use the shelter and the hay will not be trampled into the floor. Haynets are not really suitable as horses could become entangled in them.

The floor of the shelter should be raised slightly and sloping forward so that any wet and mud brought in on the horse's legs and feet, and rain which runs off the horse's back, will drain out of the shelter and keep the floor drier. The floor of the shelter can vary from being a permanent concrete base to the actual surface of the field, but this depends on the type of soil, how well it drains and the location of the shelter – for example lower-lying areas can be very boggy but may provide more shelter. Whichever surface you provide, it is a good idea to cover it quite deeply with something soft that will drain well. This will encourage the horse to lie down in a dry, sheltered place, rather than in the middle of a wet, muddy field. It is better that a clipped horse, with no protection on its belly, does not lie down in the wet because of

the risk of chills and sore, chapped skin.

If a field shelter cannot be erected for various reasons, then observe the horses to find out their favourite sheltered spot and attempt to make this area as dry as possible. A semi-permanent base could be laid down so that the horses and ponies have somewhere to stand out of the wet. This will give the legs time to dry off and heat will not be constantly lost to the cold, wet mud. A roofless windbreak could be put up; this would deflect the wind and give the horses something to shelter behind.

BEDDING

Providing adequate bedding is also very important in the management of clipped horses and ponies. To try and save money by keeping a small, thin bed is false economy, as you will usually have to compensate in other areas by providing extra feed, hay and more rugs to make up for heat loss so that the horse's condition does not deteriorate. The more coat that is removed, the more you will have to do to keep the stable, and therefore the horse, warm. Extra deep bedding is one way that you can help insulate the stable to maintain the environment at a comfortable temperature. It is very important that the horse has a good deep bed to prevent draughts which can cause a great deal of heat loss from the legs and belly. Stopping draughts along the stable floor is even more important if the legs have been clipped. When there are cold spells such as ice or snow, then extra bedding is required.

To test that the bedding is thick enough, hold a pitchfork upright about a metre/three feet above the bedding with the prongs pointing down, and then drop the fork vertically onto the bedding (make sure your feet are out of the way first!). If you can hear the prongs hit the stable floor, then the bed is far too thin. If the bedding is left up during the day, there will be a great deal of warmth lost through the floor, as this is usually made out of materials such as concrete that conduct valuable heat away. If the floor is bare, then it will quickly become wet with urine and trampled manure. This again will result in the stable becoming damp and cold. Cover as much as you can of the stable floor, at least three quarters of it, with bedding, with the sides 'banked' as this will help to insulate the walls and stop draughts.

Remove all wet and soiled bedding daily, even if it is supposed to be a 'deep-litter' system, which usually does make a very warm bed for the horse. Even if the bed is managed as deep litter, it must still must be kept clean and not allowed to become soggy. Any wet which cannot drain away through the bedding or be absorbed because the bedding is saturated, will cause irritation, soreness and possible skin infection on a horse or pony who lies down in urine without any coat on its underside for protection. Fresh bedding has to be put down regularly to maintain the depth and keep the surface dry.

There is a variety of bedding materials on the market now, each with its good and bad features. Wheat straw is still widely used and if it is of good quality will be quite absorbent and can drain urine very well. A large proportion of straw made today has been mangled through a combine harvester and is so damaged and crushed that its draining potential is reduced. Barley straw can be very irritating because of the long barley 'awns'. These can dig into the horse with thin skin and are particularly aggravating to horses who are clipped. They may put the horse off lying down altogether. Modern methods of processing tend to remove the awns so this problem is decreasing.

Shavings and shredded paper make very good bedding as both are warm and very absorbent. Droppings and wet patches should be tended to every day to keep the bed tidy, and fresh bedding put down. To some sensitive clipped horses, shavings can be irritating as they can dig into the horse when it is lying down. Shavings can also stick to the inside of rugs and bandages, so it is advisable to brush the shavings off daily. Be careful as to the source of the shavings, especially those from wood yards as preservatives are often used on the wood and a few horses may be allergic to these. Horses with clipped legs can be particularly vulnerable, so enquire if any potentially irritating substances have been added to the wood before you use it as bedding. Specialised suppliers of wood shavings for equines ensure that their product is from uncontaminated sources so there is little risk. Although very rare, some horses can be allergic to the ink off paper bedding so be suspicious if your horse has any unexplained skin conditions on the lower leg and under the belly.

A relatively new type of bedding comes from the stem of the hemp plant. The sponge-like centre of this plant is used to produce the bedding which, at first glance, looks like shavings, but is much softer to the touch and is proving to be much less irritating to a clipped or sensitive-skinned horse. It is extremely absorbent and appears to hold far more moisture than either straw or shavings; and doesn't tend to mash into 'pulp' like paper, if not managed efficiently. Because of its high absorbency, the horse will always be lying down on a fairly dry surface so the risk of chapped skin from soggy bedding is minimal. As with all other bedding materials, soiled and wet bedding must be removed daily. One manufacturer claims that the raw material for this bedding is not treated with any pesticide, fungicide, herbicide or insecticide, which is good news for 'green' campaigners. This makes it very beneficial for highly allergic horses and less likely to cause any adverse allergic reaction. It is also very warm, comfortable and insulates the stable extremely well which ultimately helps to maintain the horse's condition.

CLOTHING/RUGS

Stable rugs

Because the horse's natural insulation has been removed, artificial insulation is provided by using rugs. The type and number of rugs is governed by the amount of coat that has been removed, the weather conditions, the breed of horse or pony and whether he is susceptible to the cold. The locality and structure of the stables is another major factor in choosing rugs and blankets for the horse. If little can be done to improve a stable's position etc. then compensatory changes to the horse's management must be undertaken. It is of interest to compare the amount of clothing required by a thoroughbred, for example, with that needed by a non-thoroughbred, and a native cobby type in particular. Thoroughbreds do, of course, feel the cold, and need a lot of rugs etc. to keep them warm. But because they do not generally grow a very long, thick winter coat, what coat they do have has little warmth value in the winter, so they do not miss it so much if removed. Non-thoroughbreds on the other hand, especially those with a high proportion of native blood, usually grow a very thick winter coat for protection against the cold, wet British climate. After clipping, these horses are often seen wearing just one, relatively thin rug, which in reality does not keep them very warm as the warmth equivalent of two good quality rugs has probably been clipped off. Most people put very little clothing on a native type because they believe that these breeds are notoriously hardy. Yet the native pony may feel much colder than would a full thoroughbred given a blanket or hunter clip and a thin rug.

There are many specially designed rugs on the market these days. They are much more expensive than the old traditional jute rugs but they have many advantages. Each make has its own unique features as well as many that are common to all, so choose the one that suits your type of management and the horse's special needs. Most are very light but warm, many being made out of thermal textiles. They are also very durable and easy to machine-wash and dry. Some horses can be allergic to the 'man-made' synthetic fabrics used so be very vigilant for any signs of irritation on a clipped horse, which will usually appear around the neckline, shoulders and over the back if this area has been clipped. Washing powders used to clean rugs can also precipitate an allergic reaction. If the horse is sensitive, an untreated cotton under-rug may help.

Just as we humans need to pile on more clothes in the cold, and reduce the amount of clothing when the weather becomes warmer, so the same applies to horses. Adjust the amount of rugs according to the weather and put on extra rugs or blankets when conditions are frosty or there is snow on the ground. These extra layers can be removed gradually when the weather is warming up. Rather than having the same amount of rugs on day and night, an extra rug can be put on at night when the temperature drops so that the horse feels the benefit. The extra rug can then be removed during the day when the temperature should be warmer. If the horse is clipped early such as at the beginning of autumn, don't put a thick, heavy rug on the horse straight away as he won't feel the benefit in the winter and is in danger of overheating.

If you have a rug with a blanket lining, fleece lining or woollen under-rugs, or are using an extra blanket, then bedding (especially shavings) will stick to these and cause irritation to the skin of a clipped horse. Brush the bedding off the rug before putting it back on the horse.

Some owners rug their horses up with a fairly light rug prior to clipping so that the coat is in a better condition for the clipper blades. It is a good idea to put a light rug on a young horse for a short while so he is used to rugs before the clipping is done. I remember a young horse that was clipped, and the owners

found they couldn't get a stable rug anywhere near him! Nobody thought of trying one beforehand, not even to see if the rug was the correct fitting.

The fitting of the rug is very important and it is more economical to buy good-quality well-fitting rugs that are 'cut' right and shaped to your horse, than to muddle through with rugs that are ineffective in conserving warmth and cause damage by rubbing the horse's chest or

An extra thick, quilted stable rug with crossed surcingles.

A multi-purpose stable/travel/ sweat rug that keeps in warmth and wicks away moisture.

withers. The rug should come well in front of the withers to prevent this. If the horse is clipped, there is very little protection on the shoulders and chest, so they can be easily rubbed, especially on thin-skinned horses. These sore areas provide an easy access for infection through the skin.

Lining the wither and shoulder areas on the inside of the rug can go a long way in reducing friction. Alternatively specially designed bibs (see photo on page 92) can be purchased which fit over the withers and around the shoulders and shield this area from chafing. These bibs are a good investment if you have a horse or pony whose shoulders are prone to rubbing, even though the fitting of the rug is correct. They can be worn at night under the stable rug, and during the day under a New Zealand.

Because the belly of a clipped horse is bare and exposed, choose a rug which comes below the level of the belly as this serves to deflect any draughts. These rugs are usually described as 'deep' or 'extra deep'. If a very deep rug is used on a short-legged horse or pony then there is a slight danger of the bottom of the rug being stood on when the horse is getting up. Some makes are 'wrap around', which again helps to maintain warmth on the stomach.

A crossed surcingle is much more comfortable than a single roller that presses constantly on the same area of the back. A roller has to be much tighter than crossed surcingles, and because it resides in the same place, a single surcingle may also chafe the underneath of a clipped horse. Be very careful, though, about leaving the crossed surcingles too loose because the horse's foot may

become trapped when he tries to get up.

The back of the rug must fit snugly, and many are specially shaped for this purpose. Watch that there is no pressure on the croup or friction on the points of the hip. A rug which fits well around the back end is most important if the horse's quarters are clipped. Some rugs have a tail flap which aids heat conservation and is especially useful if the top of the tail has also been pulled.

With horses that are full, hunter or blanket clipped, the coat is clipped off the neck, and in very cold weather conditions a neck cover can be attached to the rug so valuable warmth is not lost from this area.

New Zealand rugs

These rugs are made out of a lightweight, waterproof material and are used for horses and ponies who live out in the field throughout the winter, or for clipped, stabled horses that are turned out during the day. They protect the horse very effectively from the wind and rain and are of paramount importance for a clipped horse living out.

As with stable rugs, New Zealands must be well fitting and lined with a warm material which can also be thermal. These have a looser fitting around the chest to allow the horse to lower his head to graze comfortably, but must not be so loose as to allow the rug to slip back too far. A good fit around the quarters is even more important than in a stable rug because the horse or pony always stands with his back end towards the wind and rain, which would leave the clipped areas very exposed. A tail flap and an unpulled tail will go a long way in

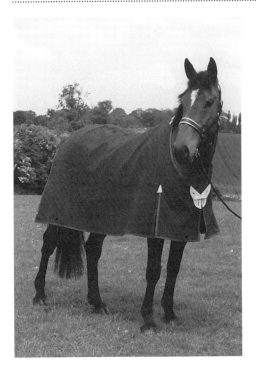

An extra deep, gusseted New Zealand rug.

A turnout rug with crossed surcingles and a tail flap for extra warmth.

reducing the amount of heat lost. New Zealands can also be bought in 'extra deep' versions and this feature is much more important than on a stable rug because it will shield the belly from the wind and prevent a great deal of warmth being lost to the elements.

Even with a very good quality rug, it may be necessary to put an under-rug or blanket on the horse in very severe weather. If he has been out in continual rain for a few days, you may have to bring him in overnight to dry him, and the rug, completely. In conditions where persistent rain is forecast over a number of days, it may be kinder to bring the horse or pony in and keep him stabled until the weather improves. This action will help save valuable condition, which may be difficult to replace with some horses. Constant wind and rain will wear a horse

down both physically and mentally. There is nothing more heart-wrenching than seeing a cold, wet horse looking thoroughly dejected and miserable. It is a good idea to have a spare rug for occasions when one becomes saturated or damaged and needs repair.

If the New Zealand is of the type that has leg straps then make sure that they don't rub the back legs where the hair has been removed. If they are made of leather, oiling them so they are softer may help considerably. The leg straps can also be threaded through a length of pliable hose pipe or the inside of a bicycle inner tube.

Chafing of the shoulders is very likely, especially for horses living out, because they have much more freedom of movement. The shoulder area on the inside of the rug should be lined, possibly with sheepskin, to prevent chafing, or a bib, as described earlier, can be worn under the rug.

Ill-fitting rugs, New Zealand or stable, can easily rub the shoulders.

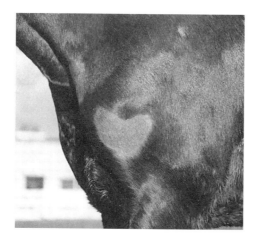

This specially designed bib prevents rug rubs and can be worn under turnout or stable rugs. It protects the shoulder, chest and withers.

An extra deep New Zealand rug with under-belly crossed surcingles and a tail flap, and a 'wrap-around' front to ensure a good fit and provide extra warmth.

Many rugs are 'self-righting', which means they correct themselves and do not hang over to one side when the horse lies down or rolls. Even so, the rug must always be taken off and adjusted at least once a day. Before replacing the New Zealand after exercise or adjustment, make sure that the rug's wither, shoulder and chest areas and the leg and belly straps have been brushed free of mud.

New Zealand rugs can be fitted with a neck cover and this is valuable for stabled horses who haven't any coat on the neck and are turned out during the day. Some horses and ponies may have been wearing a New Zealand for a while before they are brought into the stables for the duration for the winter. If an ill-fitting New Zealand has been allowed to rub the chest before the horse has been clipped, the skin may be quite sore and will need time to heal before clipping. Make sure that the stable rug is a more comfortable fit and have the New Zealand lined and the fit adjusted before it does any more damage.

Many modern materials are completely waterproof and never need reproofing – check this when you purchase the rug. The older type, canvas rugs, lose their waterproof quality with

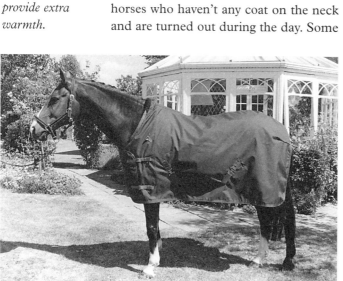

wear and eventually let the rain go straight through, especially along the seams. This can also happen if the rug is inadvertently cleaned with a detergent which breaks down the waterproofing substance. The canvas rug will certainly need reproofing, most easily achieved with an aerosol spray purchased from a saddler or camping shop. If it is just a seam that is leaking, then candle wax can be rubbed on to seal it. The company which manufactured your rug will be able to advise you on the best way of reproofing, if needed.

A New Zealand rug alone is not enough to keep any horse living out warm and in a healthy condition, especially if he is getting on in years. If he is clipped, he will need adequate shelter, extra feed and hay, increasing as the weather becomes colder and wetter.

Stable bandages

Not all horses need their legs bandaged when stabled, but in certain circumstances this can be crucial. If the horse has a full clip where the hair on the legs has been removed, then bandages are a must to protect the legs and keep them warm. In fact, for any horse who feels the cold, bandages may be necessary.

To keep the horse comfortable, bandages must be removed at least twice a day and then replaced tidily. Make sure that the legs are cleaned properly after exercise before the bandages are put back on, because any mud or sweat left under the bandages will begin to rub the legs. When the bandages are off, remove any bedding debris that may have attached itself. Gamgee is generally put underneath bandages for increased warmth.

Make sure the bandages are not put on too tightly as this will interfere with circulation. Once fastened, check the tension and readjust if too loose or too tight. Velcro is now commonly used to secure bandages, but tie tapes are still used widely. If tapes are used then remember to tie the knot on the outside of the leg.

Exercise sheets/rugs

These are smaller than normal stable rugs and are used on clipped horses when being exercised. For wet conditions, they can be made out of a light, waterproof material, which can be lined for extra warmth. There is also a variety of woollen types which are preferable if the weather is cold but dry. Horses that are full or hunter clipped will certainly benefit from wearing one of these during exercise. An exercise rug will help keep the quarters of the horse warm and prevent valuable body heat being lost from this area.

If the horse has a clip which leaves some coat over the quarters, then an

A hood and neck cover, which can be attached to a New Zealand rug, is valuable for stabled horses who haven't any coat on the neck and are turned out during the day.

A waterproof exercise sheet.

exercise rug is rarely required, but all horses are individuals, and if the horse is susceptible to the cold then a rug is necessary. Whatever the clip, if the weather is windy and extremely bitter, then using an exercise sheet is certainly much kinder. Older horses tend to feel the cold more, and for this reason may need a rug for exercise even if they still have plenty of coat.

If the horse is doing fast work or tends to jog a great deal during exercise, then they can become very sweaty underneath the rug so wearing one may be detrimental. On very mild, dry, sunny days the rug may not be required at all and it is a good idea to leave it off so that the horse benefits from the warm sun on his back. The rug is usually fitted over the quarters and under the saddle with the front folded up under the saddle flaps to make it secure. Always make sure that the fillet string is fitted correctly so that in windy conditions the rug does not blow up over the horse's back and frighten him.

Exercise sheets need not be used on clipped field-kept horses and ponies as they will have plenty of coat remaining. If rain is forecast, however, a light, unlined waterproof rug can be useful. This will keep the back dry and the New Zealand rug can then be put back on the horse as soon as you return to the yard or field, providing he is not too hot and sweaty.

DIET, COAT AND CONDITION

The horse's overall health is due to his genetic make-up, past and present environment and his management in that environment. One of the most fundamental 'environmental' factors is nutrition and this is a facet which horse owners and trainers have the most notable ability to manage (or mis-manage).

Diet is obviously determined according to work requirements, management, breed, temperament, the individual's sensitivity and reaction to feed products and, of course, the extent of the clip. We need to ensure that the diet is adequate, well balanced and 'mimics' the horse's natural diet as much as possible with regard to ingredients and administration. It pays to remember that the horse's digestive system has evolved over millions of years. Horses are 'trickle feeders', which means that they need small amounts of food continually passing through their digestive system. In his natural habitat, the horse is usually grazing for as long as it takes to fulfil his nutritional requirements, which is usually around sixteen hours depending on the quality of the forage available. Horses are not designed to utilise large concentrate feeds nor to go for long periods without food as is so often seen with stabled horses.

In-depth diet formulation is beyond the scope of this book, but a few basic pointers with regard to the nutritional needs of horses, and clipped equines in particular, are outlined here. To provide an adequate balanced diet it is important to know what the horse actually does with the food. Horses and ponies need food for a variety of reasons. A very important one, especially in winter, is to maintain body temperature and this is vital for all the bodily functions to work effectively. Food is also used to maintain weight and body condition, renew and replace body tissue and provide the energy requirements for all the horse's movements, from just wandering around the field to strenuous competition. This includes unconscious movements of the internal body systems, such as respiration, digestion etc. If adequate food is not provided, horses will start to use their fat reserves to compensate so that most of the above functions can continue. If this is allowed to persist, the horse's condition will eventually deteriorate to a stage where he will become emaciated and die. A horse needs to have an insulating layer of fat to help keep out the cold, so if a horse is already undernourished and then clipped, it will be extremely difficult to prevent him going downhill even more. Care must be taken not to let the horse get into this state in the first place and to adjust the diet before things go this far.

A thick coat can be deceiving, making the horse appear to be in better condition than he actually is. Take time to feel the horse through the coat so you really get a good idea of his true condition. Unless the horse is still competing, hunting or racing, try to let the horse go into the winter with a good covering of fat but not so much that he is grossly overweight. Obesity is unhealthy and puts unnecessary strain on limbs and vital organs, such as heart and lungs, whenever the horse is worked hard. Clipped horses, more so than unclipped ones, will utilise more of their energy reserve to compensate for the cold, so this must be considered when working out their dietary requirements. A constant supply of good quality hay has a much more 'warming' effect than concentrated feeds, and its digestion produces enough body warmth to maintain temperature. Horses and ponies can get cold for no other reason than that they are hungry. Ensure that an adequate supply of quality hay is available at all times through the winter for both stabled and field-kept horses and ponies. This will provide the continual bulk required to keep the gut working properly, assist with the digestion of concentrate feeds and help keep the horse warm.

Dietary factors may influence a horse's response to the clippers. Some individuals may be intolerant or allergic to certain feedstuffs which can affect their temperament and behaviour, making them overly 'sharp' and nervous. The intolerance may create an imbalance in the gut micro-organisms which, among other things, can lead to a heightened 'startle reflex', i.e. the horse is exaggeratedly jumpy (problems with eyesight and hearing can also cause this). If the horse is not overfed and is getting enough work but still remains a little silly while being ridden and handled generally, suspect the feeding. Keep this possibility of food sensitivity in mind if the horse or pony's behaviour has changed after a recent change of diet, especially to a different brand of feed.

When clipping, one cannot expect a quality finish on a horse with a dry, brittle coat that is full of scurf. A horse in excellent condition, will always have a smooth, shiny finish to the clip. If the coat on an unclipped horse looks bright, then you can guarantee the coat underneath will be in good condition and gleaming, even after clipping. The skin should be supple and move freely over the ribs, for example, and the hair should be fairly soft. This varies with breed and type, but it certainly should not be brittle or coarse. Scurfiness is often a sign of 'toxic' overload, or due to a lack of certain minerals or essential oil. If the horse does have scurf (dandruff), then check the mineral and trace element levels of the diet and the essential fatty acid content. Ensure that the horse is not allergic to something in his diet, to synthetic tack, or to rug materials or whatever they are washed in.

The condition of the skin, coat and indeed hooves, tends in many respects to reflect the overall condition of the horse. A dull or staring coat, long wispy hairs, failure to moult or continual moulting, scurfiness and skin lesions may reflect covert disease conditions, e.g. due to infections, parasite infestations or nutritional deficiencies or imbalances. Whilst poor skin and coat condition will always indicate health problems, it cannot be assumed that a shiny coat means that the horse is totally free of health problems. The horse's condition must be assessed as a whole, based on observation, drop-pings, thirst, appetite, urine, eyes, temperature, temperamental changes, and attitude to work.

If the horse is healthy and sweats freely, this may affect the decision to clip. If he doesn't sweat properly this may be due to some kind of hormonal or nutritional problem. As a result he may have a tendency to overheat, so don't be fooled into thinking that because your horse is not sweating, he is coping with his work. Sweating is, after all, a good safety mechanism.

Clipping can help horses who tend to 'break out' into a cold sweat after work. These types can be prone to chills, the prevention of which can be very time-consuming after a strenuous day. This 'breaking out' is a tendency of horses receiving too much protein in the diet, and it can be corrected once the diet has been adjusted. Horses who produce foamy, sticky sweat also are generally being fed too much protein.

Always consult your vet if skin and coat problems are evident. He may be able to give appropriate treatment if there is a bacterial or fungal cause or run tests for any underlying nutritional problems. He may feel it necessary to refer you to a reputable independent equine nutritionist to formulate a balanced diet for your horse.

In the wild, the horse instinctively knows what grasses and herbs he needs to maintain optimum health. When the wild horse is unwell, he is able to seek out mineral-rich herbs to help fight disease and restore balance. Because we have 'domesticated' the horse, we restrict his ability to browse for specific plants so we need artificially to provide all he needs to make up the deficit. This isn't always easy, even with all the nutritional advice available today.

All horses are individuals, so any theoretically based diet can only be aimed at a happy medium. Diets must be fine-tuned to the individual horse, even more so with competition horses, racehorses and horses and ponies who have dietary related conditions such as laminitis, azoturia and malabsorption problems. In general, the more stress the horse is subjected to, whether it be work, growth, exposure to disease, travelling, climate, clipping, psychological pressure (e.g. bullying by people or other horses), the more finely tuned the diet needs to be. Pay attention also to the teeth and plan an effective worming programme. If worm infestation or dental problems are present then the horse will not be getting full value from the feed he is receiving. Many horses will benefit from good quality supplements even if they do appear to look well.

GROOMING

Grooming not only improves a horse's appearance but is essential to health and condition. This applies to all horses whether they are stabled or living out in the field. Grooming removes mud, sweat and other waste products from the surface of the skin. If the pores are allowed to become clogged, infections can result. Horses should be groomed every day to stimulate the skin, tone the muscles and improve circulation.

Horses and ponies thoroughly enjoy a good roll in the mud and often become covered from nose to tail. This is not done to annoy their owners, as is gener-

ally thought, but is a basic survival behaviour which helps to keep them warm. If you are planning to ride a muddy horse, the dirt must be removed from all areas that are at risk of being rubbed by tack, as this will make the skin very sore. The clipped areas are particularly vulnerable because there isn't any coat left to offer protection, so be very meticulous in keeping clipped parts of the horse as mud-free as possible. The clipped areas which need attention on a field-kept horse are the shoulders, chest, under the belly, and between the front and back legs. The head should also be groomed properly before riding because the bridle will rub muddy areas. Any mud over the withers must be brushed off daily as the rug will easily rub here.

Be careful of over-grooming, though, because the horse needs to build up a certain amount of grease to waterproof the coat. Grooming helps to spread the natural oil in the coat (sebum) and distribute it evenly throughout the coat. Stabled horses also need thorough grooming to remove excess grease, sweat if ridden, and mud if they have been turned out during the day.

When grooming a clipped horse, don't remove the rug completely and let him stand around becoming cold. Keep a rug on the horse and fold back a section of it to brush a small area – the nearside shoulder, for example – then cover the finished part with the rug and go on to another area. Do this until the horse has been groomed all over then put his rug on tidily and secure the straps.

Don't use a hard dandy brush on the clipped parts of the horse. Some horses may not like being brushed with this under normal circumstances, so they are bound to show some annoyance if the dandy brush is used on bare skin, so be very thoughtful and use a soft body brush.

EXERCISE AND COOLING OFF

Whilst regular exercise is important for every horse's well-being, it is essential for a stabled horse who may be confined for long periods. Horses who lack exercise can become unmanageable and a problem to handle and ride; this can turn out to be very dangerous indeed. With adequate daily exercise and correct feeding, problems are less likely to occur and the horse's behaviour should not get out of hand.

Riding a horse who becomes sweaty every day and then having to wait hours for him to dry off so he can be fed, watered and have his rugs put back on, is very impractical as time can be very limited for most owners. It can also can be very unhealthy for the horse. Clipping will certainly make things much easier and reduce the time spent trying to dry the horse off, but this in turn does have its problems and some thought should be given to potential difficulties.

Common sense must prevail as to the amount of work the horse is asked to do, which should be adjusted according to his fitness, management, the amount of coat that is remaining and the length of time you have to make him comfortable after riding.

Horses that have just been clipped are usually much more 'on their toes' when ridden out for the first few times, so be prepared for a little excitement! They

should settle down fairly quickly with regular riding. When hacking out on a clipped horse, be careful not to stand around allowing the horse to get cold. Resist the temptation to stop and engage in 'horsey talk' when you meet someone else while out riding. Keep moving all the time, even if he is wearing an exercise rug. You'll be surprised how quickly horses and ponies can become cold. If competing in the winter, and your horse is clipped (as he should be for hard work) then take rugs with you for putting over him between classes.

Take care not to let a field-kept horse or pony with a low trace clip or with even less coat removed, become saturated with sweat. These horses must be given steady, gentle exercise so that the amount of hair removed is enough to keep them fairly cool. If worked hard and fast, no amount of clipping will prevent sweating.

To conserve warmth, (dry) horses out in the field will seem to 'fluff' up their coats. This traps warm air between the hairs and insulates the horse. Because the hairs on a horse with a full winter coat are quite long, this method is very effective in conserving a great deal of warmth. This fluffing is achieved via tiny erector muscles at the hair roots. When an impulse from the brain is received, the muscles contract, which causes the hairs to stand up. Horses who are wearing rugs or who are wet from rain or sweat are unable to employ this mechanism over a large area of their body, although rugged horses will have artificial insulation.

When a hot horse sweats, heat is conducted away from his body by the moisture passing through the skin and coat and then evaporating. The process of sweating prevents the horse overheating. A sweating horse can cool down more quickly than his coat can dry and the cooling process continues, even though the horse's normal body temperature (which is approximately 38°C/ 100.5°F) returns. If the process is allowed to continue without intervention, the wet skin and coat will cause chilling, and possibly hypothermia if the body temperature is affected. The horse's metabolic processes can also be affected, which may ultimately result in death. Because of this, no horse should ever be turned out in a field while still wet, even though he may be cool. Instead of moving around to stay warm and dry off, cold, sweaty horses and ponies will begin to shiver to keep themselves warm. This is effective for a short period but if prolonged will result in the body being depleted of energy, and the horse or pony will eventually lose condition, if he doesn't die of hypothermia first!

Clipping a horse does not mean that chilling won't happen; its prevention in only possible through correct management after exercise. If a clipped horse comes in from work and nothing is done to keep him warm, he can become cold and chilled even sooner than a non-clipped horse. On return from exercise, try to bring your horse back to the yard as cool as possible. Try and make they walk home quietly if you can; by doing this, half the job of cooling and drying is already done. If it has been raining, or if your horse is cool but still wet, then you can trot quietly for the last quarter of a mile if the terrain and environment permit. This active movement uses up energy which in its turn creates heat, serving to warm the horse up a little but not enough to make him sweat up again.

By doing this, he is not already cold when he arrives back at the stables.

As soon as your horse is untacked, put a rug on him, preferably one which allows cooling but which still keeps the horse warm – a sweat rug, which is usually made from open cotton mesh, is ideal. This alone may not be enough to keep him warm once he has cooled down so an additional rug may be needed to warm him so he dries off completely. A light stable rug can be used over the sweat rug but check that it is designed to let the skin breathe and allows moisture to evaporate – some modern materials don't have these qualities. Nowadays there are multi-purpose rugs which can be used as a stable, travel and sweat rug, and these are well worth investing in.

Never put a New Zealand rug over a wet or sweating horse as the moisture will not be able to evaporate, the horse will not dry off properly and there is a good chance he will become cold and chilled. Walk the horse around and keep him moving. Be careful if there is even a light breeze, as the air movement from this can take heat away too rapidly. If it is windy, the horse can be put in a sheltered area or in his stable, which must be well ventilated but not draughty as this can also cause problems with rapid heat loss. Feel the horse's ears and body frequently to make sure that he is warm. Groom the sweat out of the coat as soon as he is dry enough – the stimulating effect of the brush will also help to warm him. Put on his stable rug and secure the buckles and straps. Put the bedding down on return if it is not down already. This is better than letting him stand around on a bare, cold floor. He can then have a good roll if he wants to.

ROUGHING OFF A CLIPPED HORSE

Roughing off simply means gradually conditioning a stabled horse to living out in the field and hardening him up to withstand the elements without losing weight and becoming ill. This should be done very gradually, taking up to five or six weeks depending on the weather and individual horse. Thoroughbreds and Arabs will always take a little longer to adjust than will a hardy native pony. Older horses may need a full six weeks to acclimatise as they tend to feel the cold more as the years progress.

A horse should never be in racing condition one day and thrown out in the field the next, as is sometimes seen. Work and feed should be adjusted accordingly, not just stopped completely when the horse has finished hunting or racing. Reduce the concentrated feed and increase the amount of hay gradually over a number of weeks. This will help the horse's digestive system adjust to the change in living environment. If you are unsure how to how to do this consult your vet or an equine nutritionist who can advise you on a roughing-off programme for your individual horse. The amount and thickness of the rugs should be reduced in stages, again according to the weather and temperature. On sunny spring days, leave a thin rug on while the horse is in the stable then progress to leaving the rug off completely. Do the same at night when the temperatures are becoming milder; put a light rug on for a while then leave the horse overnight without one, so he gets used to this. Do not turn out for good until the horse is comfortable without any rugs when stabled at night and when turned out during the day.

Introducing the horse to the fresh spring grass must be done very carefully indeed, to prevent laminitis and colic, and this is especially important for native ponies. Laminitis is all too common and many cases could be prevented with careful management and attention to diet. Turn out for an hour a day, possibly less at first, then after a week, turn out for a little longer and increase the duration over the third week. The introduction to grass will not be as drastic for the horse who has been lucky enough to have a few hours out in the field each day during the winter months, even though the quality of grazing is poor.

Spring nights can still be very cold for horses who have been stabled throughout the winter. Even if they have been carefully roughed off before turning out for good, they may still need a light waterproof rug for a while, just to take the 'edge' off the cold night temperature and the spring showers until the summer is well and truly here.

Don't groom the roughed-off horse so much, especially with the body brush. Allow the coat to build up a fair amount of grease as protection against the cold and showers. If the weather does start off warm and dry when you begin your roughing-off programme, there is always a possibility (especially in Britain) that the conditions may take a turn for the worse. Don't let the horse suffer any discomfort; it is no disgrace to go back a step so that he is kept warm and happy. His coat, condition and behaviour will give you a very good indication of how he is coping with the change in management and living out. If his coat becomes dull and staring, and if he loses condition and is looking fed up and miserable then you can guarantee that he is not coping. This is the only way he can tell you that something is wrong, so do listen to him. If the above signs are evident, please do something about it before it is too late.

.9.

CHOOSING CLIPPERS

Before going out and buying the first set of clippers you see, take some time to think about what you expect from them and what sort of usage they will have. Also think about the price. Don't go for expensive clippers with a multitude of uses, when you only have one quiet horse and intend to clip just two or three times during the winter - this is very uneconomical. If, on the other hand, clippers are going to be purchased for a big yard with many horses at livery, or you are going to start a professional clipping service, then a hard-wearing, versatile, heavy-duty, good quality machine is essential.

If you are planning to clip a large number of horses, there will be many that are an unknown quantity, each with a different experience of being clipped, or not, as the case may be. Horses may vary from being the quiet 'rock solid' type to absolutely 'lethal', so you will need clippers that are designed to be powerful, yet quiet and manoeuvrable enough to successfully clip nervous or bad-tempered horses.

The clippers that are on the market today are far removed from the big, bulky, noisy sets that have helped to ruin many horses' attitude to clipping, and the more refined, improved designs have been welcomed with open arms by horse owners all over the world. Each make and model of clippers has its own good and bad points and it is up to you to decide which design, accessories and price are most suited to the job you have in mind. There are usually adverts for clippers in the horsey press, especially around early autumn, and magazines occasionally feature reviews of makes of clippers. Most saddlery shops stock clippers, but the range is often limited. If a make and model of clippers seems to look promising, then write to the manufacturers for details and to find out who and where your nearest stockist is. There may also be adverts in the local horse press for secondhand clippers that might be worth looking into, but always get these examined by a qualified electrician before you purchase.

Some of the features that are available are listed below, but they are not included on all types, so make note of

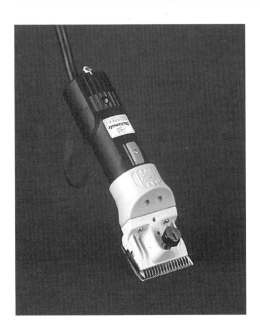

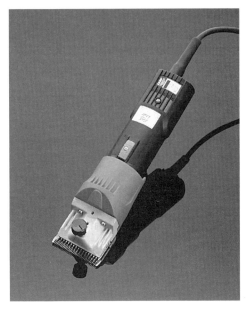

A mains powered set of clippers.

A powerful set of clippers which can be run off the mains or from a car battery.

priorities and choose the make that mostly fits your requirements and the special needs of your horse.

❍ Ask friends or acquaintances if you can have a look at their clippers, and if it is the clipping season, possibly try them out on a horse to see how you get on with them. Your friends will be able to tell you the good and bad points about their clippers from their own experiences and why they bought this particular make.

❍ Many makes are now light to handle and hold. This has come about because of the lighter materials used for the motor, and for the outside casing, which is usually lightweight plastic. This is very tough and will withstand a certain amount of ill-treatment, but care must be taken that the clippers are not dropped on hard surfaces as they are sure to

break. Some are double insulated for increased safety. Any abuse of the clippers is certainly not covered by warranty. Even though the clippers are lightweight, clipping a horse can take a long time so be prepared for aching arms, even with regular breaks.

❍ Some clippers are designed to be 'slimline'. This is very good for people with small hands and those who found the old-type clippers very bulky. It is a good idea when buying clippers to go to the suppliers and actually hold the clippers rather than choose them by mail order. Certain models have the 'right feel' and are compact and balanced. You will know whether they will be comfortable to use, and that you will be able to handle them during clipping. Many casings have a specially designed grip to make holding more comfortable and

Clippers run from a battery strapped round the operator's waist. The beauty of using this outfit is that there are no trailing leads to frighten the horse and there is no danger of the horse, handler or person clipping becoming entangled in the wires. Also you have a hand free to push the horse over and keep him still if he is fidgeting.

secure. Slimline clippers are very good for getting into the awkward places on the horse, such as between the forelegs, by the elbow and around the head.

❍ Many makes claim to be able to keep running for long periods without over-heating, and have sophisticated cooling systems built into the newer models. However, it is still advisable to stick to a rigid programme of cleaning and cooling the clippers every ten to fifteen minutes so that overheating definitely does not occur. This claim does not take into account horses who are extremely greasy and dirty, and even the best cooling system will not be effective on these horses. All makers tell you to clean and cool regularly in their operating instructions.

❍ Some designs include a 'baffle' which cools the clippers by deflecting the air over the clipper head and away from the operator. Some horses may not like the feel of this, especially those who are very difficult and sensitive. A few types have an air-cooled motor that blows air away from operator and horse, which can be better for awkward horses.

❍ Most clippers around today, even the full-sized ones, are fairly quiet (when compared to the older models) and none of the power is sacrificed to achieve this. When choosing clippers, ask to hear them running so that you can compare the noise levels of different makes. Even though they are quiet, they will still seem loud to horses because they have extremely acute hearing.

❍ Many of the old makes used to vibrate a lot and horses understandably did not like the feel of this, especially on the head. The vibration on many of the newer models is hardly noticeable, so clipping ticklish and thin-skinned horses is very much easier. Clipping the head too, will be much less of a problem with low-vibration clippers.

❍ A very versatile set of clippers is one that can be run from the mains or off a battery pack. A battery pack is very good for nervous horses and horses who tend to be bad mannered and take off around the stable without warning. There are no wires trailing around to frighten timid horses, and there is no danger of horses becoming tangled up if they do happen to run around the stable. Also, one hand is left free to keep the horse away from you if he is fidgeting.

❍ If you are really worried about getting the blade tension right, then go for the makes that have easy snap-on/off blades. Some have one-touch tension which prevents over-tightening. One possible advantage is that the blades will not have to be sent for regrinding so often as the overall clipping life of the blades is increased. Scrupulously following the

Hand clippers are useful for horses who will not tolerate the clipping machine around the head, even with gentle persuasion. Although hand clippers will get the job done, they are laborious and slow, so allow plenty if time if you have to use them. The blades require frequent cleaning and lubrication.

maker's instructions with regard to blade tension should take all the guesswork out of getting the tension right.

❍ A set of small, cordless, rechargeable clippers is very useful for finishing off tricky places, including the head and between the forelegs of narrow ponies. Some are capable of doing a whole clip providing they are fully charged. Most are fairly inexpensive and would suit the one-horse owner with plenty of time to clip his own horse – they can be quite slow. They are excellent for last-minute trimming before shows, for use around the jawline, fetlocks, mane, over the withers and for the small area of mane behind the ears where the bridle sits. They are equally good for trimming off the long cat hairs that can grow back under the jowl, down the neck, around the elbows

and around the back legs. Removing these will smarten up the horse who otherwise does not need an all-over clip. Full-size cordless, rechargeable clippers are also available on the market.

❍ A very useful feature to have on cordless clippers is the facility to plug them into the mains if the charge runs low.

❍ If mains are not available, and a cordless will not last long enough for a full clip, then the answer may be a clipping machine which has a 12-volt motor that enables it to be run off a car battery, as well as a 240 volt option.

❍ One model that I have used with a great deal of success over many years, with problem horses and in a variety of difficult situations and environments, is an extremely versatile clipper that can be used from the mains, run off a battery pack around the waist or from a car battery. By using the mains adaptor the cable is only carrying 12 volts instead of 240 volts, which is much safer.

FROM TOP TO BOTTOM: A powerful set of professional clippers; a lightweight, quiet clipper; and a cordless, rechargeable clipper.

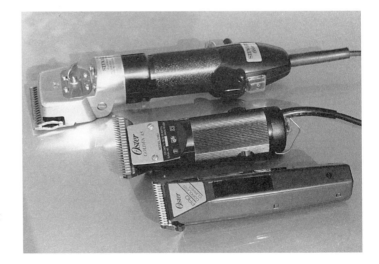

A battery-powered trimmer, smooth and quiet to run.

A cordless rechargeable trimmer, complete with finger grip.

❍ Some clippers have speed variations. They offer a high speed for horses who have a very thick winter coat, and a low speed for nervous horses and sensitive areas. A 'gadget' has been produced that looks similar to an adaptor. This accessory is designed to alter the speed of your clippers and is very useful for gradually getting the horse used to the clippers. Make sure that your clippers has an armature-type motor and your mains supply is 200-250 volts AC before using one of these.

❍ There is also a machine on the market where the motor is in a pack on a belt worn around the operator's waist. The beauty of this is that the handpiece never gets hot.

❍ Some small sets have an adjustable lever which gives a range of cutting depths. Combs can also be attached to the blades which further varies the amount of coat left on the horse. This is ideal for clipping the legs and leaving

some protection or for trimming feathers before shows in the summer.

❍ Some clippers offer a wider range of uses than just clipping horses. They can be used to trim dogs, cats and human hair. There are also clippers with interchangeable heads and blades for shearing sheep and cattle as well as horses. Go for this type if you have a variety of jobs in mind.

❍ Another useful feature to look for is the on/off switch being located on the top of the casing, just in front of where you hold the clippers. This is handy for emergencies when you need to switch off the clippers quickly when, say, a horse is panicking, has become accidentally tangled in the cables, or has been successful in squashing you against a wall.

❍ Because clippers and blades break if dropped, it is a good idea to have a loop that fits around the wrist so that the clippers do not drop onto the floor if knocked out of your hand.

❍ To prevent the motor straining and possibly seizing up, there is an overload button included on some makes which will 'pop out' to stop the clippers running. This protects the motor from serious irreversible damage. If it does pop out, make sure that the button is re-set by pushing it back in with a pen, not just your finger. Many sets are returned to the manufacturer as faulty, when the only problem is that the button has not been pushed in far enough after dealing with what originally may have caused an overload. You will hear a 'click' when the button is re-set properly.

❍ Blade designs can vary slightly, and one company has unique blades that are 'self-clearing'. With the 'self-clearing' blades, hair that would otherwise become trapped in between the blades is diverted into gaps in the top blade. This reduces the likelihood of clogging and these blades are a very good investment if you have a young, nervous or sensitive horse. There is a variety of blade thicknesses available, regulated by the number of teeth on the bottom blade (comb). They vary from very fine blades, which are commonly used by veterinary surgeons for clipping off the coat before operations or stitching wounds, to the coarse blades used for leaving some coat on the horse or for clipping sheep or cattle.

❍ If the manufacturer provides services such as regrinding blades, repairing and servicing clippers, you are assured of a complete follow-up service. If not, they should be able to give you details of a reputable company that can give a very good service if problems occur.

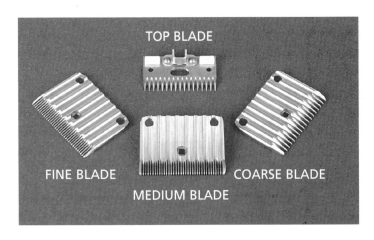

TOP BLADE

FINE BLADE

MEDIUM BLADE

COARSE BLADE

❍ Manufacturers who export clippers abroad make models that are compatible with different countries' electrical supplies and socket fittings. They will be able to give you details of overseas dealers who stock their clipping machines.

❍ Clippers and accessories usually come in handy carrying cases and the average contents are:

Clippers
Blades – usually one set of fine or
   medium blades
Cleaning brush
Clipper oil and/or grease
Blade guard
Screwdriver
Instructions

Remember to complete and return your guarantee.

Clipper blades vary from coarse to fine depending on the finish required.

BUYING CLIPPERS THROUGH A SYNDICATE

To cut down the cost of buying clippers, you can purchase a set between friends, or make up a yard syndicate. This is useful if your collective horses or ponies need frequent clips, and will work out much cheaper than if you have to hire someone to clip them each time. If there are enough people involved, you will get your money back the first time you use the clippers. Another advantage is that the clippers are readily available at the yard if you need to use them often to get a young or nervous horse used to the clippers. You should also never be stuck for someone to help you clip because they may need help in return when they have to clip their own horse.

If a group of people do buy a set, then it may be necessary to draw up some agreement as to the usage, e.g. who is responsible for the cost of the servicing, regrinding blades, any damage, or what happens if a member of the syndicate leaves the area, sells his horse or falls out with everyone else. It may be agreed, preferably in writing, that everyone is responsible for 'chipping in' towards the cost of repairs and servicing. Regrinding of the blades may be debatable, especially if one member blunts all the sets of blades in one go on a very dirty horse. This point can be overcome by each member buying a set of blades for use on his own horses, so that there is no argument about who blunted the blades. By doing this, each person knows he has a good, sharp set of blades available when it comes to clipping his own horses.

It may be an idea to make some sort of provision to 'buy out' a member who leaves, or to let someone else buy his share. The group should also agree on whether the clippers can be lent to anyone outside the syndicate. This is usually not a good idea as clippers can be damaged and arguments can arise about who is responsible – the outsider or the person who lent them out?

· 10 ·

CLIPPER CARE AND MAINTENANCE

KNOW YOUR CLIPPERS

Instruction books for your model of clippers will usually give you information on electrical specifications, basic preparation of the horse, fitting blades, tensioning of the blades, lubrication, care of blades, a 'troubleshooting' section for problems, maintenance, parts description and illustrations naming the various parts. Read the manual thoroughly before using the clippers as using them incorrectly can soon lead to problems.

Familiarise yourself with the clippers, practise setting them up, changing blades, cleaning and oiling them etc. so you are not fumbling around when you come to use them on the horse. Constantly referring back to the instructions increases the length of time it takes to clip, which can result in a cold, bored, fidgeting horse.

CARE OF YOUR CLIPPERS

Always follow the manufacturer's instructions with regard to the use and care of the clippers. This way, any problems that occur while the clippers are still under guarantee will be remedied quickly as the clippers have not been misused or abused in any way. The guarantee usually only covers defective parts, not any faults caused by accidents, negligence or improper operation. If problems occur, do not attempt to mend the clippers yourself as this will nullify your guarantee. If the instructions are followed then the clipping life of the machine is extended and the likelihood of breakdowns is reduced. Most of the care and maintenance instructions apply to every make and model of clippers, but there may be special rules that are unique to the set you have purchased, so pay special attention to these.

❍ Make sure your clippers are serviced regularly. This will ensure that all parts have been thoroughly checked, and the hair and dust that finds its way into the motor can be removed before it causes serious problems.

❍ Only use blades that are designed for

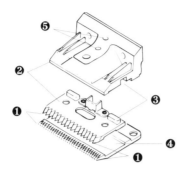

LUBRICATION POINTS:

1. *Between top and bottom blades at front.*
2. *Between top surface of top blade and underside of clipper head.*
3. *Around slot in underside of clipper head.*
4. *Between top and bottom blades at rear.*
5. *Oil-holes in clipper head.*

the same make and model as your machine.

❍ Never let the clippers overheat. When in use, cool them every ten to fifteen minutes, more frequently if the horse is even slightly dirty.

❍ Clean and oil the clippers and blades regularly during clipping, again this should be every ten to fifteen minutes or when the blades become clogged. You should always lubricate between the blades, the cutter-blade channel, crank shaft and roller – the instruction manual will tell you where these are. Don't be mean with the lubricating oil. Many new clipping machines have been returned to the manufacturer as being faulty because the motor was running slow, straining and seemingly lacking power, but all they needed was a thorough oiling. The sluggishness was caused by the blades running nearly dry. You will hear the difference in speed when they are well oiled. While in use, at least once remove the blades and give them a good clean – be careful not to lose any of the parts.

❍ Make sure that the tension of the blades is correct. If the tension screw is too tight, then the motor will sound strained, the movement of the blades is slow and excessive wear will occur to the machine and blades. The motor will quickly overheat and could even blow up if the blades are not readjusted and there is not an overload button on the machine to protect the motor. If the tension screw is too loose, then the blades will rattle and the noise will sound too fast. This can also result in blunting of the blades. The instructions supplied with the

Don't be mean with the lubricating oil. Refer to the manufacturer's instructions to find the lubrication points on the particular model you are using. Wipe off any excess oil from the blades.

Tension nut
and spring

Clipper head

Blades

Tension bolt

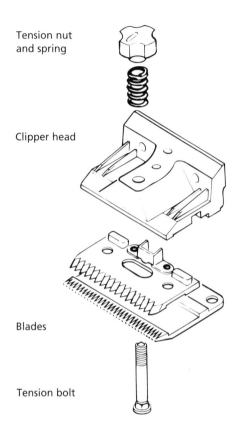

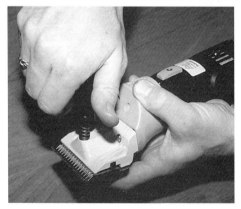

Assembling and tensioning the blades.
Following the maker's instructions when
fitting the blades will prevent all sorts
of potential problems. Most makes are
fitted with a tension bolt, spring and nut,
which secure the blades. Tighten the nut
down as far as you can, then turn it back
one to one and a half turns (or according
to the manufacturer's instructions).

clippers will tell you how to achieve the correct tension. If the blades fail to cut, even though you feel you have followed the maker's guidelines, never increase the tension, instead clean and reassemble the blades and try again.

○ Regularly brush the hair away from inside the blade head and air filters (if your model of clippers has filters). The air filters can be removed by unscrewing them, taking care not to lose the small screws. Brush away the hair from both sides of the

Air filters need to be cleaned regularly. They have to be removed from the machine as both sides can become clogged. First, remove the screw (above). Brush both sides of the gauze to remove the hair until you can see through the filter (above right). If the gauze is very dirty and greasy then it will have to be washed in detergent (right). Dry thoroughly and replace. Never run the clippers with the filters removed.

filter – removing it from just the outside is not enough to clear the gauze sufficient to stop the clippers overheating. Once removed, a high-pressure airline can be used to blast the hair away from the mesh. If very dirty and greasy and you cannot see clearly through the gauze, then it can be washed in detergent and dried before being screwed back into place on the machine. Aerosol sprays, because of their 'waxy' nature, can cause the hair to stick to the gauze filters, making them block very quickly. Unfortunately, you don't always have full control of the direction of the nozzle and the spray goes over a wider area than just the blades.

❍ When finished clipping, give the machine and blades a thorough cleaning before putting away. The clipper head can be removed so that you can clean the hair and dirt away from the gears. Lubricate the gears with the grease recommended by the manufacturer as this is especially made for their product. Some makes of grease are not designed for use on electrical equipment and can conduct electricity to the clipper blades, which is not good for electrically sensitive horses.

❍ The blades can be washed in dilute antiseptic and/or disinfectant solution to prevent spreading any possible infection to another horse. Look to see if any of the teeth have been broken. Throw away any damaged blades.

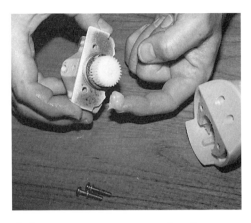

At least once a year – more often if you can – remove the head, clean the hair and dirt from the gears and apply a grease that has been recommended by the manufacturer. Some types of grease can conduct electricity through the clipper to the blades, which is not good for horses who are very sensitive to electricity.

Use a ballpoint pen or something with a suitable point to push in the overload button if it has popped out. Before continuing, deal with the problem that has caused the overload, e.g. check that the blades are not clogged, are running dry or that the filter isn't blocked.

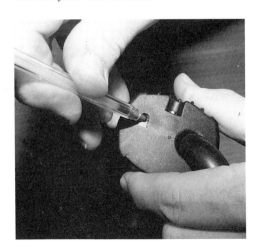

○ If blunted, send the blades away for regrinding. Always store sets separately in pairs because they are manufactured and ground to work as a matched pair. Have corresponding numbers engraved on the top and bottom so you know which are pairs – it is so easy to get them mixed up.

○ Apply a thin layer of oil or grease to the blades to stop them from rusting while in storage. They can be wrapped in grease-proof paper, an oily rag or in their original plastic case.

○ Keep the clippers in a dry, safe place. It is important that the clippers are not damp before they are put away for storage, because the metal parts that cannot be protected by grease will rust. Grease the clipper head so that it does not corrode whilst stored. Most modern clipper heads are made out of non-rust material or are coated to protect them against rust. Run the clippers until they are warm and by doing this any moisture will evaporate. Put the clippers away immediately, providing they are not too hot.

○ Check the cables for any damage, especially if the horse has been naughty and trampled on them a few times. Store the clippers and cables in a container where they cannot be chewed by rats or mice. Also, check the plug for any signs of scorch marks.

SERVICING CLIPPERS

This should be done after each clipping season has finished, not left until the autumn. Again, you will find adverts for servicing electrical equipment, but it is sensible to send the clippers back to the company who manufactured them, if they have a servicing department. This way you know that they have experienced personnel and that any parts that have to be replaced will be readily available. In theory the clippers will be returned quite quickly, correctly serviced with the proper parts fitted. If the company does not have a servicing department, then they will be able to recommend to you a reputable and reliable company who knows their products well.

Be aware of your entitlements if under guarantee. Quotes for servicing should be obtained beforehand, preferably in writing, prior to the clippers being dispatched. Have some proof that the clippers have been sent, just in case of any loss. It is also wise to send a written request, that apart from just the servicing, if any part does need replacing, you are notified first, especially if the part is expensive. If you have a very old set, sometimes it is more economical to buy a new set rather than have your old machine repaired. If one part is badly worn, you can rest assured that other parts are on their way out, so keep this in mind before agreeing to expensive repairs. As well as correct care of the clipping machine, regular servicing helps to extend the life of the clippers.

REGRINDING BLADES

The clipping life of the blades, before they have to be reground, varies enormously. New blades can retain a good cutting edge and give a smooth, quality clip for up to ten horses, sometimes

Blades with teeth missing should never be used – the danger of cutting the horse is increased and tell-tale lines are always left after clipping.

more. The number of horses that can be clipped before the blades become blunt depends largely on the management of the horses and how the clippers are handled. You need to consider the style of clips, the breed of horse, how clean the horses are, if the blades have been cleaned and lubricated enough during clipping, and if the tension of the blades has been correct. Again, manufacturers may recommend a company that offers this service. If not, look for adverts in the horsey press. Sometimes blade-regrinding services are advertised in local saddlers and feed merchants. Always send blades to be sharpened immediately after the clipping season has finished. By doing this, you will miss the rush in September or early October.

Always send blades to a company or independent person who has a good reputation and provides a quality service. Blades have to be ground to a very fine precision finish and with a good company the quality of this finish is ensured. If a set of blades are returned that are not up to standard, and do not cut the coat cleanly, then most reputable companies will usually regrind them free of charge.

Enquire if this is offered before sending the blades.

Blades are expensive to buy and you don't even want to lose one, let alone four or five, so it is wise to obtain proof that they have been sent. Don't send all the blades that you have at once; put them on 'rota', which means that you have some blades to use while others are being sharpened.

The teeth of the blades are fragile and can easily break if knocked or dropped. The steel blades are especially hardened for good cutting qualities and so are more brittle than unhardened steel. Package them very carefully with plenty of padding to cushion any impact in transit. Sometimes the blades are returned coated with preserving oil which has to be removed before being used on the horse. Wash them in blade wash and dry thoroughly before lubricating with oil recommended by manufacturer.

There are blade-sharpening kits on the market, which means you don't have to keep sending blades away and that sharp blades will always be available to use. If you have many horses to clip and a lot of blades, buying one of these can save you money in the long run, but it may not be economical for the twice-a-year clipper with one horse.

BLADE WASHES AND AEROSOL SPRAYS

Blade washes are excellent for flushing out the grease and hair that accumulates between the top and bottom blades. Some are medicated and it is claimed that they prevent clipper rash and sterilise the blades to reduce the risk of cross-infection. New blades have preservatives coated on them by the manufacturer to prevent rusting. This can initially prevent the blades cutting cleanly through the coat so it has to be removed before using on the horse. Put the blades in the wash for a while then wipe off the excess before fitting on to the clippers. The washes offer some degree of lubrication for the blades, but clipper oil should be always be used also. It is advisable to follow the clipper maker's instructions as to the use of blades washes and oils. Most manufacturers make blades washes, oils and grease as part of their clipper range. Check in the instructions if they recommend the use of blade washes for their product. There is always a good reason if they don't, so follow their advice.

When using the blade wash, make sure that only the depth of the blades is immersed and not any part of the clipper housing. Be careful that you do not let the clippers tilt back so that the blade wash is allowed to run back over, or into the clippers. This can get into the motor, which is not good for the clippers. Don't put all the blade wash into the container, a depth of two inches is quite enough.

Blade washes are expensive, especially if you have many horses to clip. When cleaning the clipper blades, loose hair and grease gets in the blade wash in the container, so it usually has to be thrown away after being used only once. A way of re-cycling the blade wash is to filter it a few times through a fine meshed cloth over a funnel, the end of which is inserted into a bottle. This way the blade wash can be used a few times before it has to be discarded. Keep the filtered blade wash in a separate container from the original unused wash. Methylated spirits can also be used for washing the clipping blades. Remember that some sensitive horses can be allergic to blade washes and oils. Try putting some on a small area of skin a few days before clipping and watch for any reaction.

There are a number of cooling, lubricating sprays available which claim to cool and clean the blades without leaving a heavy residue of oil. Although cooled and cleaned, the blades will certainly still need oiling thoroughly after and throughout clipping. Some makes of aerosols leave the blades nearly dry because the very thin oil evaporates, so very little lubrication gets between the blades. Certain brands are keeping up with the 'green issue' and are even labelled 'ozone friendly – CFC free'. Even so, only use sprays where there is plenty of ventilation. Prolonged inhalation and contact with the skin may cause irritation.

Blade washes are very harmful if swallowed, so be very careful about leaving them around, especially if there are any small children at the stables. Clearly label the containers and keep them in a safe place, preferably under lock and key.

·11·

ELECTRICAL SAFETY

Electric shock

Safety with electric clippers is very important, not only for the person clipping but especially for the horse. The body's nervous system controls conscious and unconscious movement, both in animals and humans. Muscles are stimulated by electrical impulses from the brain which are electrochemical in nature. These signals are only a few millivolts, which is very small. When an external, more powerful current is introduced to the body's internal circuit, normal function is disrupted and this is known as 'electric shock'. This abnormal current affects the muscle, and even someone who is conscious is unable to control his muscular movement. If a person was holding a live cable, he would be unable to release it because of the muscular spasm caused by the electric shock.

Animals are far more sensitive to electricity and suffer far more trauma and damage from being electrocuted than do humans. Death can occur at a much lower voltage.

Circuit-breakers

Because of the serious effect that electric shock can have on horses and humans, circuit-breakers are essential to prevent any injuries and possible fatalities. Detection of high current can be by one of two methods and many circuit-breakers employ both. The first method is by thermal detection, which is very slow; here the excess heat that is brought about by the high current expands a bi-metal strip which trips the breaker. The second method is the quicker, more efficient

Using a circuit-breaker is always wise as an extra precaution against electric shocks.

magnetic type, wherein the coil carrying the current creates a magnetic field. This serves to attract an iron part when the current is too great and this then trips the breaker.

When using a circuit-breaker, don't rely solely on this to prevent serious injury. Because it is also electrical, it too can malfunction, although this is very rare. So, do not become complacent with other aspects of electrical safety.

SAFETY CHECKS

❍ Check that your installations comply with IEE (Institute of Electrical Engineers) Regulations.

❍ Make sure that the clippers are **serviced** regularly.

❍ Have the whole system checked by a **qualified electrician**, especially if there is any short-circuiting. Never attempt to fiddle with the motor, this should be the domain of the people trained to do so.

❍ If mains operated, check that everything is **properly earthed**. Refer to the manufacturer's instructions to find out about earthing requirements.

❍ Always use a **circuit-breaker** socket or plug.

❍ Make sure that the **cables** are not worn and frayed and are not damaged from being trampled by a horse when used previously.

❍ See that the **plug** is in good condition and that the correct amp fuse is used. If fuses keep blowing for no obvious reason, this can be a sign of dangerous wiring.

❍ If, when in use, the **motor** sounds odd, turn off immediately and do not use again until the clippers have been thoroughly checked.

❍ Switch off the clippers at the mains as well as the on/off switch when not in use. Never leave them unsupervised. Disconnect from mains supply when cleaning, changing blades or dismantling clippers. Remove plugs carefully and don't ever remove them by pulling the flex.

❍ Be wary of hot plugs and sockets. Don't use a socket with **brown scorch marks**; check the plug for these marks also.

❍ If using an extension cable that is in an enclosed case, make sure that is **fully unwound** before the clippers are switched on. If this is not done, an electrical field can build up around the cable.

❍ Check that there are **no kinks** or knots in the cable.

❍ Don't let the horse stand on the cables – try to **keep them off the floor** at all times. If a horse does tread on cables and the inside wires are damaged and exposed, change the cable and have the damaged one repaired by a qualified electrician. To prevent the horse trampling the cables run it through loops of string or over non-metal, blunt hooks to keep it off the floor.

❍ Some horses love to **chew** anything, and cables are no exception!

❍ Clip on a **dry day**, even slight drizzle is potentially dangerous.

❍ Don't ever stand in a puddle.

❍ Wear footwear with **rubber soles**. This precaution also applies to the person holding the horse and anyone who may be involved with the clipping.

❍ Remove all stable fittings that are potential conductors of electricity. This means any metal objects and water buckets. Automatic-filled water containers should be emptied and covered.

❍ Rubber gloves are handy as a precaution if problems arise, but by using a circuit-breaker the need for these is reduced. These may only be necessary as a last resort if the circuit-breaker malfunctions.

❍ Know where the **fire extinguishers** are kept and if they can be used on electrical equipment. Familiarise yourself with the procedure for using them as any delay in putting out fires could be potentially tragic and at the very least costly .

❍ Have **rubber matting** on the floor. Some large yards have specialised clipping and utility boxes that are equipped with rubber floors for this purpose.

❍ To gain height stand on a wooden or plastic block – never anything metal. Don't use a pallet as your feet could get stuck between the gaps.

❍ If the floor is wet (ideally it should be dry), and the horse does happen accidentally to step on the cable and breaks the insulation, the wires will be exposed, so be prepared for 'sparks'. This is why using a low voltage set of clippers and a circuit-breaker are essential. If the horse soils or stales, clean this up before proceeding.

❍ Never clip a wet horse; the risk of shocks is greatly increased.

•QUICK REFERENCE•
TO PROBLEMS AND SOLUTIONS

LINES OBSERVED ON HORSE

Reason	What to do
Next 'sweep' of blades on coat not sufficiently overlapping first.	Allow more overlap.
Uneven pressure on horse.	More practice needed; clippers not held firmly enough.
Teeth missing from blades.	Discard and replace with new.
Edges of blades too sharp, seem to 'scratch' horse.	Reduced with certain types of blades.

UNABLE TO OBTAIN SMOOTH FINISH

Reason	What to do
Blades becoming blunt.	Change blades.
Uneven pressure of blades on horse.	Practice needed and/or adjust hold of clippers.
Very dirty, greasy horse.	Clean before clipping – wash or groom thoroughly.

BLADES FAIL TO CUT

Reason	What to do
Blades blunt.	Replace with new, sharp blades.
Hair trapped between blades.	Remove and clean blades.
Tension too loose.	Re-tension.
Tension too tight.	Re-tension.
Horse's coat dirty and greasy.	Groom thoroughly or bath.
Horse's coat tangled or matted.	Groom thoroughly before clipping.
Horse's coat damp.	Make sure horse is dry beforehand.
No oil between blade surfaces.	Make sure blades are well lubricated.

BLADES BECOMING BLUNT QUICKLY

Reason	What to do
Being used on dirty horses.	Make sure horse is clean before clipping.
Not using enough oil during clipping.	Lubricate blades frequently during clipping.
Hair often becoming clogged between teeth.	Clean frequently during clipping.
Tension too tight.	Re-tension.
Tension too loose.	Re-tension.
Inappropriate blades (e.g. fine blades used on thick coarse coat.)	Change blades.

MACHINE BECOMING VERY HOT

Reason	What to do
Filter/air vents blocked with hair and dirt.	Check filter every 10-15 minutes.
Hand over air vents.	Adjust holding position.
Blades too tight.	Re-tension.
Lack of oil between blades.	Lubricate often.
Operator running machine for too long.	Allow to cool every 15 minutes.
Armature not aligned properly.	Repair by qualified person.
Gears dry and seizing up.	Lubricate thoroughly.

MOTOR RUNNING SLOW AND LABOURED

Reason	What to do
Carbon brushes worn.	Replace by manufacturer.
Blades too tight.	Re-tension.
Lack of oil.	Lubricate thoroughly.
Dirt and hair between blades.	Clean blades frequently to prevent this.
Partially seized gearbox.	Intervention by expert.
Worn bearings.	Replace by manufacturer.

CLIPPING MACHINE WILL NOT SWITCH ON

Reason

Broken switch.

Blown fuse.

Disconnected wire.

Mains failure.

Circuit-breaker cut-out.

Overload button has been activated
(if model used has one).

What to do

Replace by manufacturer.

Replace with correct amp.

Reconnect or solder by electrician.

Check supply.

Find reason for cut-out.

Find reason for overload (see below)
and reset.

MACHINE STOPS

Reason

Blown fuse.

Overload button has been activated
(if model used has one).

Switch breaks while clipping.

Power cut.

Circuit-breaker cut-out.

What to do

Replace.

(a) Blades clogged.
(b) Filter blocked.
(c) No oil on blades.
(d) Electrical fault.

Repair by manufacturer.

Check supply.

Find fault before continuing.

INTERMITTENT STOPPING AND STARTING

Reason	What to do
Loose wire.	Check plug and cable.
Loose wire in clipping machine itself.	Repair by electrician.
Faulty switch on machine.	Replace by manufacturer.
Faulty supply.	Check mains.
Faulty plug.	Check plug.
Worn brushes.	Replace by manufacturer.
Severed cable inside covering.	Mend or replace.
Severed cable inside leads.	Mend or replace.

•USEFUL ADDRESSES•

CLIPPERS

**Lister Shearing
Equipment Ltd**
Dursley
Gloucestershire
GL11 4HR

Tel: 01453 544830
Fax: 01453 545110

Stockshop Wolseley Ltd
Lodge Trading Estate
Broadclyst
Exeter
EX5 3BS

Tel: 01392 460077
Fax: 01392 460966

Weatherbeeta Ltd
7 Riverside
Tramway Estate
Banbury
Oxon
OX16 8TU

Tel: 01295 268123
Fax: 01295 296036
*(Distributors of Heiniger,
Wahl and Sunbeam-Oster
clippers.)*

Brookwick Ward
88 Westlaw Place
Whitehill Estate
Glenrothes
Fife
KY6 2RZ

Tel: 01592 630052
Fax: 01592 630109
*(Distributors of Rex,
Hauptner and Sunbeam-
Oster clippers.)*

Cox Surgical Ltd
Edward Road
Coulsdon
Surrey
CR5 2XA

Tel: 0181 668 2131
Fax: 0181 668 4196
*(Distributors of Heiniger
and Wahl clippers.)*

RUGS AND CLOTHING

Weatherbeeta Ltd
7 Riverside
Tramway Estate
Banbury
Oxon
OX16 8TU

Tel: 01295 268123
Fax: 01295 296036

Aerborn Equestrian Ltd
Pegasus House
198 Sneinton Dale
Nottingham
NG2 4HJ

Tel: 0115 9505631
Fax: 0115 9483273

Bossy's Bibs
6 Beech Walk
Tring
Herts
HP23 5JQ

Tel: 01442 824033

COMPLEMENTARY THERAPIES

BACH FLOWER REMEDIES	RADIONICS	HOMOEOPATHY
The Dr Edward Bach Centre Mount Vernon Bakers Lane Sotwell Wallingford Oxon OX10 0PZ	**The Radionic Association** Baerlein House Goose Green Deddington Banbury Oxon OX15 OSZ	**The British Association of Homoeopathic Veterinary Surgeons** Alternative Veterinary Medicine Centre Chinham House Stanford in the Vale Oxon SN7 8NQ
Tel: 01491 834678 Fax: 01491 825022	Tel: 01869 338852	Tel: 01367 710324 Fax: 01367 718243 *NB: All written enquiries should be accompanied by an SAE.*

PUBLICATIONS

Equine F.A.C.T.S. Ltd
52a Springfield Road
Linslade
Leighton Buzzard
Bedfordshire
LU7 7QS

Tel: 01525 381373
Fax: 01525 382848

This is an invaluable reference source of complementary medicines available to equines. An annual directory, it includes names and addresses of practising complementary practitioners, including veterinary surgeons and basic information covering a variety of therapies.

•INDEX•